Prions

Contributions to Microbiology

Vol. 7

Series Editor   *Axel Schmidt*, Wuppertal

Basel · Freiburg · Paris · London · New York ·
New Delhi · Bangkok · Singapore · Tokyo · Sydney

....

Prions

A Challenge for Science, Medicine and Public Health System

Volume Editors   *Holger F. Rabenau*, Frankfurt/Main
Jindrich Cinatl, Frankfurt/Main
Hans Wilhelm Doerr, Frankfurt/Main

17 figures and 19 tables, 2001

KARGER   Basel · Freiburg · Paris · London · New York ·
New Delhi · Bangkok · Singapore · Tokyo · Sydney

Contributions to Microbiology

formerly 'Concepts in Immunopathology' and
'Contributions to Microbiology and Immunology'

....

Priv.-Doz. Dr. Holger F. Rabenau

Institut für Medizinische Virologie
der J.W. Goethe-Universität Frankfurt/Main
Paul-Ehrlich-Strasse 40
D–60596 Frankfurt/Main (Germany)

Library of Congress Cataloging-in-Publication Data

Prions: a challenge for science, medicine, and public health system / volume editors,
Holger F. Rabenau, Jindrich Cinatl, Hans Wilhelm Doerr.
   p. cm. – (Contributions to microbiology, ISSN 1420-9519; vol. 7)
Includes bibliographical references and index.
ISBN 3805571240 (alk. paper)
1. Prion diseases. 2. Prions. I. Rabenau, Holger F. II. Cinatl, Jindrich. III. Doerr, Hans
Wilhelm. IV. Series.
QR201.P737.P756 2000
616.8–dc21

00-050676

Bibliographic Indices. This publication is listed in bibliographic services, including Current Contents® and Index Medicus.

© Copyright 2001 by S. Karger AG, P.O. Box, CH–4009 Basel (Switzerland)
www.karger.com
Printed in Switzerland on acid-free paper by Reinhardt Druck, Basel
ISSN 1420–9519
ISBN 3–8055–7124–0

Contents

Preface

After the great successes of modern hygiene and medical microbiology, many people thought the threat of infectious diseases had disappeared. However, emerging and re-emerging infections with more or less pathogenic potential disproved this opinion. The onset of the AIDS epidemic has brought the old problem of *slow virus diseases* to general attention. For many years, it had been postulated that some of those diseases, particularly of the brain, were caused by unconventional viruses. Infections with these agents do not induce classical inflammations – acute or chronic encephalitis due to a specific immune reaction – but a slow and irreversible degeneration of the central nervous system presenting as encephalopathy. Those diseases had been considered rare events in animals and humans, only interesting medical and veterinary doctors and scientists – scrapie in sheep and Creutzfeldt-Jakob disease (CJD) in man were known only to specialists. The onset of the bovine spongiform encephalopathy (BSE) epidemic in Great Britain completely changed the situation. The fatal risk of consuming contaminated beef became a topic. And indeed, the fears were materialized when the first cases of new variant CJD (vCJD) were identified.

From the scientific point of view, it was a real sensation, when infectious agents without any nucleic-acid-based genome were hypothesized, which were later called proteinaceous infectious organisms (prions). The dogma that self-replicating biologic agents depend on genomic information conserved in the sequence of nucleic acids had to be given up. However, it should not be forgotten that prions are real 'viruses' (in the original sense of toxic material), as defined by the very early virologists. A lot of questions arose: How do those agents replicate? Do they always harm the host? Which factors of virulence can be identified? What are the mechanisms of pathogenicity? How are the agents transmitted? What are the options of early laboratory diagnosis and development of therapy and prevention?

In this edition of *Contributions to Microbiology, leading scientists in different fields of biomedical research show once again how important it is to deal with the prion 'problem' in a multidisciplinary manner.* The long-lasting discussion about the virus or prion theory is getting a clearer background. Furthermore, the structures of prions and their molecular biological analysis, and the questions of strain variations and species barrier are discussed. Another chapter tells us what we can learn about prions from yeast experiments. The possibilities of inactivation and disinfection of prions are of great importance for public health. This includes the test methods and their problems as well as the recommendations for clinical use and official regulations.

Another part of this book concentrates on the causes and clinical diagnostic aspects of TSE. While previously vCJD could be diagnosed only postmortem, new tests now allow identification of the disease with some confidence while victims are still alive (at least 1 year before its clinical onset). The tests include tonsil and appendix biopsies and magnetic resonance imaging. Furthermore, the epidemiology of human and animal prion diseases, disease management and the risks to public health or biologists (e.g. in the pharmaceutical industry) are also discussed.

In summary, this book informs about the state of the art of prion infection and disease.

Frankfurt/Main, May 2000

Holger F. Rabenau
Jindrich Cinatl
Hans Wilhelm Doerr

Rabenau HF, Cinatl J, Doerr HW (eds): Prions. A Challenge for Science, Medicine and Public Health System. Contrib Microb. Basel, Karger, 2001, vol 7, pp 1–6

Transmissible Spongiform Encephalopathies: The Viral Concept and an Application

Heino Diringer

Robert Koch Institute, Berlin, Germany

TSEs – The Virus Concept

Until about 15 years ago, transmissible spongiform encephalopathies (TSEs) were of interest only to a few scientists who ventured into research on an infectious, virus-like agent with unconventional properties. For example, this agent does not elicit a measurable immune response and can only be detected and quantitated in long-lasting laborious infectivity studies using animals. Of course, sheep breeders were interested in eradicating scrapie in sheep, but, as there was no evidence that this disease of sheep could ever pose a threat to human health, there was relatively little interest in TSEs amongst doctors. A disease like Creutzfeldt Jakob disease (CJD) with only 1 case in a million annually, caused by a similar unconventional agent in elderly patients was not the most important medical task to pursue. Even when AIDS was discovered and it was recognized that a slow disease caused by a lentivirus posed a threat to human health, the interest in TSEs remained surprisingly low although scrapie and visna maedi in sheep, the latter caused by a lentivirus, were the diseases that led to the concept of slow-virus diseases [1].

It was the new formulation [2] of an old speculation that raised new interest in TSEs. It proposed that only a protein without an associated nucleic acid could transmit an infectious disease, explain strain variation, and replicate identically to enormously high titres. Apparent experimental evidence this time seemed to support the concept [2] and the agent was given the name 'prion'. Although the experimental evidence was invalid [3], this time the concept raised curiosity and particularly attracted young scientists inexperienced in infectious diseases but well trained in molecular biology techniques.

The prion concept gained uncontrolled scientific preference and – after the appearance of bovine spongiform encephalopathy (BSE) in cattle and subsequently new variant CJD (vCJD) in humans – led to the award of the Nobel prize for the discovery of the infectious agent.

And yet it is questionable [4] whether this option and the statement that fractions containing the prion almost entirely consist of the prion protein [5] are correct. Thus, it is likely that the agent causing TSEs has not yet been discovered, and models that propose an undiscovered virus-like agent as the cause for TSEs therefore are still very important.

In this article, firstly, I will shortly present one such virus-based concept that was developed during the last 15 years and will cite some examples of its usefulness in the past. Secondly, I will apply virological thinking to find an explanation for the predominant appearance of vCJD in unusually young patients.

A Worst-Case Scenario

In 1988 an epidemiological study was published on a new form of a TSE, i.e. BSE [6], affecting cattle, a species used for human food consumption. The term 'a worst-case scenario' suddenly was brought up, implicating the possibility of a transmission of BSE to humans. But a worst-case scenario in terms of virology takes much more into account:

For example, a virologist would like to find out and discuss at what time the disease actually appeared in cattle, he would like to estimate the likelihood of transmission to humans, find out the most likely route of transmission, the source of transmission and he has to think of possibilities of further spread in the human population.

The agent causing BSE retains its properties when transmitted to other animal species and humans [7]. It thus appears to have a broad range of hosts and seems to be much more virulent than other TSE strains. A virologist confronted with the transmission of such an agent in humans cannot overlook the fact that at least in scrapie there is good evidence for a preference for a maternal transmission of TSE. There are good indicators that this may also be true for BSE in cattle. The vehicle for the agent that would explain this preference for a maternal transmission is not understood yet.

Humans use drugs and blood transfusion, and thus spread in humans could be different from natural spread in animals.

There are the aspects of the sensitivity of currently available tests to detect infectious materials.

There is the enigma of the very young age of patients affected by vCJD.

The answers to all these questions are intimately associated with the nature of the agent and thus with the question whether we indeed know the agent or still have to search for it.

The Viral Concept and Its Advantages in the Past

The viral concept to understand TSEs was put forward in 1985 [8] and simply describes these diseases as virus-induced amyloidoses.

On the one hand, it postulates that TSEs are caused by a yet undiscovered virus-like agent that spreads through the organism like conventional viruses (see, for example, the spread to the CNS [9]) Thus, the hypothesis uses rules of general virology to explain the epidemiology of TSEs [10] and to evaluate possible risks of transmission [11, 12].

On the other hand, the viral concept, from a biochemical point of view, predicts that the pathogenesis of these TSEs must be understood as an amyloidosis; in other words, the putative virus causes the aggregation of a host cell protein to form a classical amyloid fibril. In 1985, this was a very new pathogenetic mechanism in virology and a new concept to understand TSEs. It describes TSEs as 'hidden amyloidoses' [8, 13] and thus opened the possibility to think of other degenerative diseases of the CNS as amyloidoses and correctly postulated the existence of other 'hidden amyloidoses'. The concept was put forward before it was known that the disease-specific amyloid fibril protein was indeed derived from a host protein, a prerequisite for understanding amyloidoses or for describing these diseases from a chemical point of view with Glenner [14] as β-fibrilloses. When it turned out that the amyloidogenic protein was a glycoprotein of the surface of neurones, the virus model was extended to the description of an interaction of the agent with its cellular receptor to understand cell tropism.

In contrast to the prion concept in the past, this model was very efficient in foresight. For example it stressed the importance of the oral route as the main natural route of transmission [10, 12], the association of animal and human TSEs [10, 12], the likelihood of transmission of BSE to humans [10, 12], the genetic susceptibility of the human population [11, 15, 16]. Apart from these applications, I have described the concept many times in more detail [11–13, 16–18] as well as its differences compared to the prion concept [11, 16]. I will now apply virological thinking only to one unsolved problem: the unusually young age of patients affected by vCJD.

Table 1. Mean age of kuru cases

Mean age of patients years	Number of patients		Ratio female/male patients
a	b	a × b	
7.5	68	510	1.5
12.5	75	938	1.3
17.5	40	700	1.5
25	29	725	8.4
35	12	420	28

Boxed cases (183 cases 2,148 years) were used to determine the mean age of children and adolescent cases. These young kuru cases have a female to male ratio of about 1.5 as in the case of vCJD. Mean age: 12 ± 4 years From Gajdusek [21]. (If all patients in the table were included, the mean age would be 15 ± 8 years.)

Why Are Most of the Patients Suffering from vCJD so Young?

I will present evidence that the young age of onset of vCJD may indicate infection in infancy. Remember the experimental evidence that vCJD is caused by transmission of the BSE agent to humans [7], most probably by food. Oral transmission usually requires a high load of infectivity as found in tissues such as the spinal cord or brain rather than in any other tissue. The source of food that may have contained such material is unknown.

vCJD affects young adults, mostly under 35 years, in contrast to sporadic CJD, which affects elderly adults. The reason for this is unknown. The ratio of females to males affected by vCJD is 1.5. The onset of disease in the 40 cases of vCJD in the UK is assumed to have occurred between 1994 and 1997. Apart from 9 cases with an age of onset over 35 years, 31 cases developed vCJD at the age of 35 or younger, including 25 patients even younger than 30 years and 8 patients even younger than 20 years. The mean age of these 31 patients was calculated to be 25 ± 6 years (mean $\pm$ SD).

The vCJD is the only example with good epidemiological and experimental evidence of transmission of a spongiform encephalopathy (TSE) from animals to humans [7], i. e. across a species barrier. When passing to a new species, the incubation period for a given route of inoculation is usually increased not only for conventional viruses but also for agents causing TSEs.

For example, the primary and secondary transmissions of the BSE agent to VM mice result in incubation periods of 433 and 116 days, respectively [19]. The transmissions of BSE and vCJD to transgenic mice containing the gene for the human TSE-specific amyloid result in incubation periods of 602 and 228 days, respectively [20].

Kuru is the only example of transmission of a TSE in humans in ritual cannibalism. Transmission occurred mostly in adult women together with their very young children of both sexes exposed to infected brain material containing a TSE agent already adapted to humans. In kuru, transmission was by the peripheral route and may well have been oral. I have calculated a mean incubation period of 12 years for those patients, who must have been exposed to the kuru agent in their very early infancy [21] (table 1). In my extrapolation I have included only children and adolescents, with a female to male ratio of about 1.5. This ratio of 1.5 is similar to the ratio found in patients with vCJD and one can surely assume that all or most of these kuru patients were infected in infancy or early childhood. Among older patients, the female to male ratio is higher because males did not attend the ceremonies.

If for vCJD we similarly assume an exposure to infection in early infancy, the mean incubation period is 25 years. This is twice the mean incubation period I have calculated for kuru. Such an increase in the incubation period for an agent crossing a species barrier must be expected in view of the data presented for BSE [19, 20] and vCJD [20].

Exposure of patients with vCJD to the agent of BSE in early infancy could be an explanation for the appearance of vCJD predominantly in younger adults. However, such a possibility would imply an exposure of infants to the BSE agent already in the late 1960s and 1970s, before the production of meat and bone meal was changed [6] and before BSE was recognized.

Such argumentation is in agreement with the concept that BSE may have been the result of a transmission of scrapie from sheep to cattle on meadows in the UK [10] which then, as a slow viral disease, spread silently until its existence was finally recognized by the increased fortuitous spread through feed containing infectious meat and bone meal. It is important to follow up such concept by epidemiological studies in order to understand the appearance of BSE and CJD, for which, according to the prion concept, a stochastic misfolding of Prp^C to Prp^{sc} was the decisive event.

References

1 Sigurdsson B: Rida. A chronic encephalitis of sheep. With general remarks on infections which develop slowly and some of their special characteristics. Br Vet J 1954;110:341–354.
2 Prusiner SB: Novel proteinaceous infectious particles cause scrapie. Science 1982;216:136–144.
3 Diringer H, Kimberlin RH: Infectious scrapie agent is apparently not as small as recent claims suggest. Biosci Rep 1983;3:563–568.
4 Diringer H, Beekes M, Özel M, Simon D, Queck I, Cordone F, Pocchiari M, Ironside JAW: Highly infectious purified preparations of disease-specific amyloid of transmissible spongiform encephalopathies are not devoid of nucleic acids of viral size. Intervirology 1997;40:238–246.
5 Prusiner SB: The prion diseases. Sci Am 1995;January:30–37.
6 Wilesmith JW, Wells GAD, Cranwell MP, Ryan JBM: Bovine spongiform encephalopathy. Epidemiological studies. Vet Rec 1988;123:638–644.
7 Bruce ME, Will RG, Ironside JW, McConnell I, Drummond D, Suttle A, McCardle L Chree A, Hope J, Birkett C, Couzens S, Fraser H, Bostock CJ: Transmissions to mice indicate that 'new variant' is caused by the BSE agent. Nature 1997;389:498–501.
8 Diringer H, Braig RH, Pocchiari M, Bode L: Scrapie-associated fibrils in the pathogenesis of diseases caused by unconventional slow viruses; in Magistretti PJ (ed): Discussions in Neurosciences. Geneva, Fondation pour l'Etude du Système Nerveux, 1986, vol III: Molecular Mechanisms of Pathogenesis of Central Nervous System Disorders, pp 95–100.
9 Beekes M, McBride M, Baldauf E: Cerebral targeting indicates vagal spread of infection in hamsters fed with scrapie. J Gen Virol 1998;79:601–607.
10 Diringer H: Proposed link between transmissible spongiform encephalopathies of man and animals. Lancet 1995;346:1208–1210.
11 Diringer H: Bovine spongiform encephalopathy (BSE) and public health; in Aggett PJ, Kuiper HA (eds): Risk Assessment in the Food Chain of Children. Nestlé Nutrition Workshop Series Vevey, Nestlé/Philadelphia, Lippincott/Baltimore, Williams & Wilkins, 1999, vol 44, pp 225–233.
12 Diringer H, Beekes M, Oberdieck U: The nature of the scrapie agent: The virus theory. Ann NY Acad Sci 1994;724:246–258.
13 Diringer H: Hidden amyloidoses. Expt Clin Immunogenet 1992;9:212–229.
14 Glenner G: Amyloid deposits and amyloidosis: β-Fibrilloses. N Engl J Med 1980;302:1333–1343.
15 Diringer H: Creutzfeldt-Jakob disease. Lancet 1996;347:1332–1333.
16 Diringer H: The development of the viral concept on transmissible spongiform encephalopathies (TSEs) and its application to issues concerning public health; in Zichichi A, Goebel K (eds): The Science and Culture Series. International Seminar on Nuclear War and Planetary Emergencies. Singapore, World Scientific, 1997, vol 22, pp 85–94.
17 Diringer H, Beekes M, Baldauf E. Cassens S, Özel M: Amyloidosis: The key to the epidemiology and pathogenesis of transmissible spongiform encephalopathies; in Gibbs CJ Jr (ed): Bovine Spongiform Encephalopathy – The BSE Dilemma. New York, Springer, 1996, pp 251–270.
18 Diringer H, Özel M: Übertragbare spongiforme Enzephalopathien – wodurch werden sie verursacht? Spektrum der Wissenschaft 1997;3:74–76.
19 Fraser H, Bruce ME, Chree A, McConell I, Wells GAH: Transmission of bovine spongifom encephalopathy and scrapie to mice. J Gen Virol 1992;73:1891–1897.
20 Hill AF, Desbruslais M, Joiner S, Sidle KCL, Gowland I, Collinge J: The same prion strain causes nvCJD and BSE. Nature 1997;389:448–450.
21 Gajdusek DC: Infectious amyloides: Subacute spongiform encephalopathies as transmissible amyloidoses; in Fields BN, Knipe DM, Howley PM (eds): Fields Virology, ed 3. Philadelphia, Lippincott-Raven, 1996, pp 2851–2900.

Prof. Dr. Heino Diringer, Ladestrasse 48, D–26180 Rastede (Germany)
Tel. and Fax +49 4402 3134, E-Mail heino-diringer@planet-interkom.de

Rabenau HF, Cinatl J, Doerr HW (eds): Prions. A Challenge for Science,
Medicine and Public Health System. Contrib Microb. Basel, Karger, 2001, vol 7, pp 7–20

·······················

The Prion Theory: Background and Basic Information

Detlev Riesner

Institut für Physikalische Biologie, Heinrich-Heine-Universität, Düsseldorf,
Deutschland

Non-Virus Properties of the Scrapie Agent

As early as 1966, Alper et al. [1] suggested, based on the anomalous
resistance of the scrapie agent against both ionizing and UV irradiation, that
the target size is too small for a viral genome. Irradiation at different wave
lengths of UV light showed that scrapie infectivity was equally resistant at
250 and 280 nm [2]. Because proteins are more sensitive to UV irradiation at
280 nm than at 250 nm [3], Alper et al. did not follow up their original
idea, that the scrapie agent could be a protein. A nucleic acid genome with
unconventional features, an abnormal polysaccharide with membranes and a
purely membranous nature of the scrapie agent [4] were discussed in the
literature. In 1967, Griffith [5] had published a model of a self-replicating
protein, which, however, did not trigger further research.

Other features became known that pointed to subviral agents. For example,
no viruses or virus-like particles could be identified in electron micrographs
of highly infectious material; the simplest explanation was that the agent was
too small to be detected The target size from ionizing radiation was originally
determined to be about 150 kD [1] and later revised to 55 kD [6]. As a further
untypical feature of a virus, it was realized that the scrapie agent did not
initiate an immune response.

It was Stanley Prusiner from the University of California in San Francisco
who extended the early studies systematically and summarized his own and
other results as follows:

(1) Chemical and physical procedures which modify or destroy nucleic
acids (meaning also those of viruses, bacteria, fungi, etc.) do not deactivate
the scrapie agent.

(2) Chemical and physical procedures which modify or destroy proteins, do, however, deactivate the scrapie agent.

Nobody except Prusiner was brave enough at that time to formulate the logical conclusion from the results outlined above: the scrapie agent is not a virus but a proteinaceous infectious agent, which he called a 'prion' in his corresponding publication in *Science* [7]. This however, triggered a worldwide controversy, and some of the arguments are summarized by Diringer [8, this volume]. An infectious agent without a nucleic acid as information carrier was – and for some researchers still is – too much of a violation of a central dogma of molecular biology. It was argued that most experiments have lent only indirect support to the new agent, and one might might envisage an unidentified virus with unconventional properties that prevent its identification. The earlier data had argued that the scrapie agent was as small as a viroid, but the speculation that a viroid might be the cause of scrapie [9] was given up when the properties of viroids and those of the scrapie agent were compared systematically and found antinomic [10].

The Concept of Two Isoforms of the Prion Protein

If the agent consisted predominantly or solely of protein, the next task would be to isolate and characterize the protein. This took several years and resulted in a single protein of 33–35 kD, called the prion protein (PrP). It is glycosylated, has a C-terminal glycolipid anchor, is highly hydrophobic, and appears as a membrane-associated protein [11]. After purifying the protein, Prusiner together with L. Hood and Ch. Weissmann and their teams, identified and cloned the gene of the protein. Surprisingly or not, the gene of PrP is a single-copy gene of the host [12].

How could a host-encoded protein be at the same time harmless to the cell and act as a lethal infectious agent when invading from the outside? It was an important step in the development of the prion model, when it became evident, that the PrP from uninfected animals was sensitive against degradation by proteinase K (PK) whereas the highly purified scrapie agent was highly resistant against proteolysis [13]. This result clearly showed that the PrP was present in different conformations, states of aggregation, possibly complexes with other molecules, whatever one could imagine to explain the difference in PK resistance. Although the PrP from noninfected and infected animals was chemically identical [14] – i.e. same amino acid sequence and chemical modifications – one had to assume functional differences, so-called isoforms. Whereas the cellular isoform PrP^C is produced in the healthy organism, the abnormal or scrapie form PrP^{Sc}, or PrP^{CJD} and PrP^{BSE} is produced in the infected organism in cases of Creutzfeldt-Jakob disease (CJD) and bovine spongiform encephalopathy (BSE), respectively. PK resistance served indeed as marker of infectivity, and the term 'PrPres' was introduced to emphasize the particular feature of PK resistance. Some authors use PrPres synonymously with PrP^{Sc}, others point out that of PK resistance is a biochemical and infectivity a biological property. The situation is depicted in figure 1 of a healthy and an infected hamster.

The insolubility in aqueous solution is closely related to the PK resistance. PrP^{Sc} forms highly insoluble depositions in the infected brain which might have different forms, from compact plaques to quite diffuse depositions. As a consequence of the infection, PrP^{C} is transformed into PrP^{Sc} which is then present in a much higher concentration. It is noteworthy that the golden hamster became the experimental animal of choice, since the incubation time after intracerebral injection is fairly short (around 3 months) and which is even more important it is strictly related to the titer of the infectious material. During PK treatment of PrP^{Sc}, about 70 N-terminal amino acids are truncated, leading to the PK-resistant core of 27–30 kD, called PrP 27–30. Since infectivity is not reduced by the PK treatment, it is obvious that PrP 27–30 still possesses full infectivity and that the 70 N-terminal amino acids are not involved in the infectious process. Treatment with both PK and mild detergents like sarkosyl transforms the infectious material into an even more compact fibrillar form with amyloid properties, which are called either prion rods or scrapie-associated fibrils (SAFs). Figure 1 also shows differences in the secondary structure of PrP^{Sc} and PrP^{C}, which will be discussed in more detail in another chapter of this book [Cappai et al., pp. 32–47]. For more details of the differences between PrP^{C} and PrP^{Sc}, the reader might refer to reviews by Prusiner [15–17].

Basic Experiments Pro and Contra Prions

Three experimental approaches should be mentioned which tackle the question of an informational molecule like a nucleic acid in prions in a very direct manner. In addition, there exists a wealth of experimental results from molecular biology, cell biology, genetics which are analyzed in another chapter of this book [Cappai et al., pp. 32–47].

The PrP$^{o/o}$ mouse is a transgenic mouse in which both alleles of the *PrP* gene have been knocked out [18, 19]. If such a mouse is inoculated with prions, no disease symptoms can be observed, and the disease cannot be passaged further from this animal. Such a result is in complete agreement with the prion model, since it shows that PrP is responsible for the pathogenic

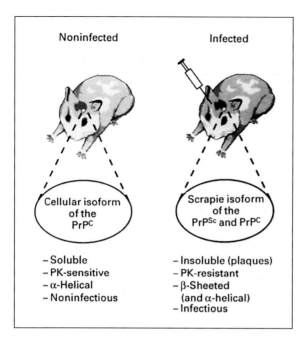

Fig. 1. Schematic presentation of PrP isoforms.

effect as well as for the infectivity. However, the results do not prove the prion model, since they only show that PrP is necessary but not that it is sufficient.

The phenomenon of scrapie strains is not so easily reconciled with the prion model. The first experiments were carried out by the Scottish scrapie researchers Bruce and Dickinson [20] and Kimberlin et al. [21]. The experiments can best be carried out with hamsters. Two different 'isolates' of infectious scrapie material – not considering the origin of different isolates – were used to inoculate genetically identical hamsters. In the first passage, a species barrier had to be overcome, and therefore the incubation time might be different after inoculation with different isolates. In the second passage, however, prions with an identical PrP sequence were used for inoculation, and still clearly different incubation times and different lesion patterns in the brain were obtained. Prions seem to contain more information than the mere sequence of PrP. This phenomenon is called 'prion strains', and several researchers take it as an indication of the presence of an informational molecule, most probably a hitherto undetected nucleic acid.

The Search for a Nucleic Acid

The existence of prion strains prompted several research groups to search systemtically for scrapie-specific nucleic acids in infectious material (for a review, see Riesner [22]. One approach was to search for differences in the nucleic acid profiles of infected versus uninfected cells or tissue. Those studies were carried out by end-labeling all nucleic acids and comparing the patterns in two-dimensional gel electrophoresis. Another way was differential hybridization screening of cDNA libraries. It was also tried to analyze nucleic acids which co-purify with scrapie infectivity. Finally, nucleic acids still present in highly purified infectious material were cloned and screened for exogenous sequences. None of the approaches led to the detection of nucleic acid other than the host's, the cloning vector or bacterial or fungal impurity. One should, however, have in mind that the fact of not finding a nucleic acid does not prove its nonexistence.

Because of all the negative results, approaches were attempted either to find or to exclude scrapie-specific nucleic acids [22–26] The concept was to start with highly purified prions and perform two types of measurement: (1) size and numbers of nucleic acid molecules present in the infectious material, and (2) number of infectious units. The ratio of the numbers of nucleic acid molecules and infectious units (P/I) might lead to two different conclusions. If the ratio were larger then unity, no definite conclusion would be possible. If, however, the ratio were smaller than unity, more infectious units would be present than nucleic acid molecules and therefore the nucleic acids would be excluded from being essential for infectivity. The numbers of infectious units were determined in a biotest using the incubation time assay; for the nucleic acid analysis a novel method, return refocussing gel electrophoresis, had to be developed. As outlined in the original literature, this technique was adapted to the very peculiar situation, i.e. that nothing was known about the nucleic acid to be analyzed, and DNA or RNA, double- or single-stranded and even heterogeneous nucleic acid had to be taken into consideration. As a summary of several experiments, the P/I ratio, depending on the length of the nucleic found in the infectious material is shown in figure 2. Only nucleic acids smaller than 80 nucleotides (50 in recent as yet unpublished studies) are present in amounts with P/I >1, i.e. could still be essential for scrapie infectivity; larger nucleic acids, i.e. also those of viruses or viroids, are excluded. Thus, the nucleic acid analysis represents a completely independent proof that the agent of scrapie cannot be a virus.

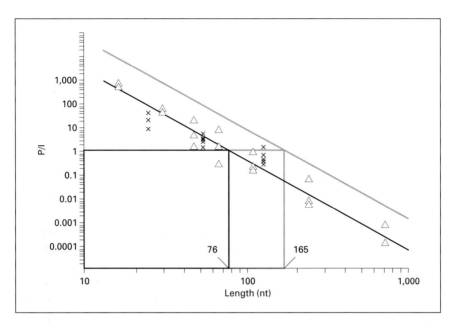

Fig. 2. Ratio P/I from five independent prion samples. Only for small nucleic acids (<76 nt) is the ratio P/I above unity. The calculation is based on the assumption that the hypothetical scrapie-specific nucleic acid is a well-defined molecular species among the heterogeneous nucleic acids. Data were taken from Meyer et al. [23] (×), and Kellings et al. [25] (Δ). The straight line is an interpolation in order to determine the average of the length of the nucleic acids at P/I of unity. The upper line represents a conservative estimation of the experimental errors.

Structural Transitions of the Prion Protein as Basis for the Replication Mechanism

The structures of PrP^C and PrP^{Sc} are presently the basis for any models of prion replication. Only schematic pictures are used here; molecular structures and models are described in the next chapter. In the noninfected organism, PrP^C is expressed in concentrations varying from tissue to tissue, and is located on the outer membrane of the cell and anchored by its glycolipid anchor. It is degraded with a relatively high turnover rate. In the course of an infection, one might assume that the invading PrP^{Sc} comes into close contact with PrP^C, and due to this contact a conformational transition of PrP^C to PrP^{Sc} is induced. Such a mechanism as proposed by Prusiner's group is, depicted schematically in figure 3. The intermediate state of PrP^C and PrP^{Sc} being in contact is called a heterodimer; this is the reason why PrP^{Sc} alone has been assumed to be present as a dimer. Induced conformational changes of proteins are known from chaperone actions, and in this sense PrP might act as its own chaperone. The transition of PrP^C to PrP^{Sc} might go on as long as new PrP^C molecules are synthesized in the cell. The newly generated PrP^{Sc} will form new aggregates, or be incorporated into already existing aggregates; it will be stabilized by aggregation, thereby acquiring a much longer turnover time. The size and internal structures of PrP^{Sc}-aggregates vary from species to species and disease to disease as revealed by comparison of the different CJD disease variants (see Collinge et al. [27] and Knight et al. [28], this volume].

It is presently not possible to perform the conversion of PrP^C into PrP^{Sc} in the test tube. Such an in vitro conversion would be the ultimate proof of the prion model which is still lacking. In the following, a brief outline will be given on the features of conformational transitions underlying the PrP^C to PrP^{Sc} conversion and on different models of replication.

Besides studies on fragments of PrP and on recombinant PrP, experiments have been carried out on three natural forms of PrP, i.e. PrP^C, PrP^{Sc} and the N-terminal-truncated but still infectious form of PrP^{Sc}, called PrP 27–30. Biochemical and particularly biophysical experiments suffered from the fact that PrP^{Sc} and PrP 27–30 are insoluble in aqueous buffers and even mild detergents. Consequently, strong effort has been made to solubilize infectious prions, or biophysical methods were adapted to insoluble samples like in thin films or applying spectroscopy with microbeams.

As a summary of solubilization experiments, it became evident that prions lost their infectivity whenever they were solubilized. PrP^{Sc} or PrP 27–30 was either denatured, as for example in guanidine hydrochloride , or urea [29, 30], or its structure was shifted to a more PrP^C-like structure as in low concentrations of sarkosyl (2%) or sodium dodecylsulfate (SDS, 0.2%) [31].

Nevertheless, the different biophysical approaches led to a more complete picture of the different structures of PrP, as summarized in table 1. The molecular structure of PrP^C was derived from an NMR analysis [32, 33].

Several approaches to in vitro conversion of PrP^C into PrP^{Sc} have been reported in the literature. Caughey et al. [34] incubated radiolabeled PrP^C with a large excess of PrP^{Sc} under partially denaturing conditions [36]. They could show that PrP^C acquired PK resistance after renaturation, and could induce even specific N-terminal truncation sites. However, because of the large excess of PrP^{Sc}, it could not be tested whether infectivity was newly generated; thus possibly PrPres but not PrP^{Sc} was induced. Later it was shown by utilizing strain-specific features (see chapter by Collinge et al. [27] that infectivity was not generated [37]. Kaneko et al. [38] induced a transition from a soluble PrP^C into an insoluble, PK resistant structure by a large excess of the PrP peptide 90–145. PrP 27–30 which was solubilized by SDS could be retransformed into

Table 1. Properties of PrP^C and PrP^{Sc}/PrP 27–30

PrP^C	PrP^{Sc}/PrP 27–30
Soluble	insoluble, present in different forms of aggregates
PK-sensitive	PK-resistant
3 α-helices of 14–25 amino acids, small antiparallel β-sheet of 2 × 3 amino acids[1]	β-sheet formation (up to 30%) in addition/or partially in addition to the α-helices[2]
Noninfectious	Major or only component of infectivity

[1] The molecular structure of PrP^C in solution is known from an NMR-analysis [32, 33] with the structured C-terminal part (amino acid 110–231) and a more or less structureless N-terminal part.

[2] For PrP^{Sc}/PrP 27–30 secondary structure contents were estimated from CD and JR measurements [34, 35]; only rough structural models but no molecular structure are available.

an insoluble state by adding 25% acetonitrile [31] or by mere removal of the SDS [39]. The secondary structure, too, was changed back to the characteristic β-sheeted structure of prions and partial PK resistance was induced. However, acquired infectivity as a consequence of the conformational change, was not found [39].

As a summary of the studies reported above, one has to conclude that the properties of β-sheeted structure, insolubility, and PK resistance are found in prions but are not strictly correlated with infectivity. PK resistance could be induced by several methods without, however, acquiring infectivity. PK resistance was correlated with aggregation and aggregation with β-structure, but these features are not sufficient for infectivity [30, 31, 40]. At present, it cannot be decided whether the right conditions for induced infectivity have not been found, a second, not yet identified component is still missing, or a principal feature of the PrP^C→PrP^{Sc} transition is not yet understood. It should be noted that a rigid chemical analysis of highly purified prion rods showed small amounts of specific lipids [41] and significant amounts, i.e. above 10% of a polyglucose scaffold [42]. It is, however, not clear whether these components are essential for infectivity.

Alternative models of prion replication have been reported in the literature by Prusiner et al. [43] and Come et al. [44]. Both are models of PrP conformational transitions and both are hypothetical in the sense that firstly they are derived from the 'protein-only hypothesis' and secondly assume conformational transitions occurring in PrP-free in solution.

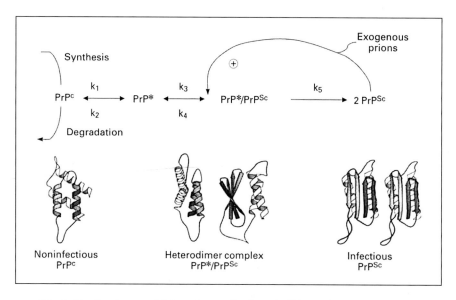

Fig. 3. The Prusiner model for prion replication: PrP^C is synthesized and degraded as part of the normal metabolism. PrP* is an intermediate in the formation of PrP^{Sc}, which can revert to PrP^C or be degraded prior to its binding in PrP^{Sc}. In infectious prion diseases, exogeneous prions enter the cell and stimulate conversion of PrP* into PrP^{Sc}. One should note that the molecular structures depicted are from model calculations, not from NMR analyses. The figure is taken from Huang et al. [46].

As depicted in figure 3, Prusiner et al. [43] proposed the 'heterodimer model'. A complex of PrP^C and PrP^{Sc} is formed, in which PrP^C is transformed into PrP^{Sc} similar to the enzymatic mechanism of an induced fit. More recently, the model was specified in that a protein X has to be involved in the complex [45]. Regarding an equilibrium, $PrP^C \leftrightarrow PrP^{Sc}$, PrP^{Sc} would be the favored state, otherwise there would be no driving force for the catalytic turnover which is reversible at equilibrium.

Another mechanism was proposed by Lansbury and his colleagues [44], in which fibril formation – as known for actin or β-amyloid and called 'linear crystals' – and generation of infectivity are closely connected. The model is depicted in figure 4. Monomeric PrP^C is in fast equilibrium with a PrP^{Sc}-like conformation, with PrP^C being the favorable state. A number of PrP^{Sc}-like molecules can form aggregates with decreasing concentrations, down to a nucleus of n PrP^{Sc}-like molecules. If the nucleus has formed, growth of the aggregates is faster than dissociation, and increasingly larger aggregates will be formed. In that mechanism – and in contrast with the heterodimer-model the first stable aggregate corresponding to a functional PrP^{Sc} will be the nucleus of n PrP^{Sc}-like molecules.

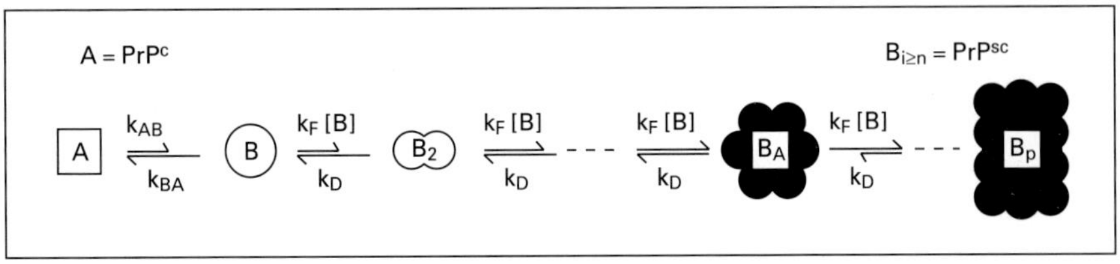

Fig. 4. The Lansbury model for prion replication PrP^C and the PrP^{Sc}-like structure are in fast equilibrium, with PrP^C being the favorable state. Several intermediates have to be formed before a stable nucleus of n PrP^{Sc}-molecules is generated. The stable nucleus is the first state with the typical prion properties. The figure is modified from Eigen [47].

Eigen [47] has treated both models according to the exact law of chemical thermodynamics and kinetics. He concluded that the 'heterodimer model', which is in essence a linear autocatalysis, cannot be adapted to realistic thermodynamic and kinetic constants. Within the time span of a mammal's life an infection would either never happen or would always happen, in clear contrast to reality. If, however, several PrP^{Sc} molecules had to cooperate to transform one PrP^C molecule, a mechanism which would be similar to an allosteric enzyme action, threshold effects would come into play and such a mechanism would not contradict well-established thermodynamics and kinetics. With this extension, the Prusiner and the Lansbury models show some similarity in the sense that several PrP^{Sc} molecules are involved in the conversion process.

Infectious, Sporadic and Familial Etiology of Prion Diseases Based on the Prion Model

If the PrP^C→PrP^{Sc} transition is the primary event of the disease, essential conclusions can be drawn or else, remarkable experimental findings might be explained on the basis of this assumption. Firstly, a conformational transition which is catalyzed by the contact with other proteins or factors would also occur spontaneously according to the laws of thermodynamics, although the probability and the rate might be very low. If only a single or a few PrP^C molecules undergo the transition, they could start the autocatalytic cycle, and from the cell where it happened an infectious process could spread through the whole body or organ. This could be an explanation for the sporadic manifestations of CJD described in later chapters of this book [Knight et al., pp. 68–92; Zerr et al., pp. 93–104]. Secondly, if a protein has two or more structural alternatives, both of which are stable, or one of which is metastable, one should expect germline mutations which favor PrP^{Sc} over PrP^C. Favoring need not be limited to PrP^{Sc}, but could also be a higher incidence for the spontaneous PrP^C→PrP^{Sc} transition. Indeed, many prion diseases with hereditary manifestations are known, and all are connected with mutations in the PrP gene (see reviews by Prusiner [11, 48] and chapters on CJD and other human prion diseases in this volume [Knight et al., pp. 68–92; Zerr et al. pp. 93–104]). Finally, the sporadic as well as the familial forms of the disease were transmitted to experimental animals – or in the case of medical accidents to humans. These findings support the general concept of prion diseases. In summary, prion diseases are exceptional also in the sense that they can have an infectious as well as a sporadic or familial etiology. Thus, it is a great merit of the prion concept as well as convincing support for it that diseases of quite different etiologies can be reduced to a unique phenomenon.

Outlook

Since the transition of PrP^C to PrP^{Sc} cannot yet be performed in the test tube, biophysical results on the structure and structural transitions of PrP will be rated as high, as they might lead to the desired in vitro PrP^C→PrP^{Sc} transition which then would represent the ultimate proof of the prion model. It will not matter whether a physical, chemical, enzymatic or any other treatment of PrP^C is applied as long as infectivity is newly generated from a noninfectious PrP^C sample. At present it is not clear whether the conformation of PrP^{Sc} is already present in the monomeric PrP, in an oligomeric state or how big the smallest size of the infectious entity might be. Infectious prions are only available as large aggregates; is this an artifact from the preparation or an intrinsic feature of the infectious state? The potential hidden difficulty is evident from the finding that about 10^5 PrP molecules are needed for one infectious unit. Are they all identical in conformation, and is the incidence of infection so low? Or do much fewer PrP molecules exist which are the really infectious PrP^{Sc} molecules which are protected by 10^5 PK-resistant PrP molecules? How could the large aggregates be dissociated to generate more infectious particles? – an essential question, particularly with respect to the Lansbury model? Do the lipids and the polysaccharides found in prions play an essential role for the infectious process, either by stabilizing the PrP^{Sc} conformation or by inducing contacts with the cell surface? As mentioned above, most experiments and models refer to conditions in solution; it might be that the transition works only in close contact with the membrane, i.e. closer simulation of cellular conditions would be required for a successful PrP^C→PrP^{Sc} transition.

Acknowledgment

The author is grateful to H. Gruber for her help in preparing the manuscript.

References

1   Alper T, Haig DA, Clarke MC: The exceptionally small size of the scrapie agent. Biochem Biophys Res Commun 1966;22:278–284.
2   Alper T, Cramp WA, Haig DA, Clarke MC: Does the agent of scrapie replicate without nucleic acids? Nature 1967;214:764–766.
3   Setlow R, Doyle B: The action of monochromatic ultraviolet light on proteins. Biochim Biophys Acta 1957;24:27–41.
4   Hunter GD, Kimberlin RH, Gibbons RA: A modified membrane hypothesis. J Theor Biol 1968; 20:355–357.
5   Griffith JS: Self-replication and scrapie. Nature 1967;215:1043–1044.
6   Bellinger-Kawahara C, Kempner E, Groth D, Gabizon R, Prusiner SB: Scrapie prion liposomes and rods exhibit target sizes of 55,000 Da. Virology 1988;165:537–541.
7   Prusiner SB: Novel proteinaceous infectious particles cause scrapie. Science 1982;216:136–144.
8   Diringer H: Transmissible spongiform encephalopathies: The viral concept and an application; in Rabenau HF, Cinatl J, Doerr HW (eds): Prions. A Challenge for Science, Medicine and Public Health System. Contrib Microbiol. Basel, Karger, 2001, vol 7, pp 1–6.
9   Diener TO: Is the scrapie agent a viroid? Nature 1972;235:218–219.
10   Diener TO, McKinley MP, Prusiner SB: Viroids and prions. Proc Natl Acad. Sci USA 1982;79: 5220–5224.
11   Prusiner SB: Scrapie prions. Annu Rev Microbiol 1989;43:345–374.
12   Oesch B, Westaway D, Wälchli M, McKinley MP, Kent SB, Aebersold R, Barry RA, Tempst P, Teplow DG, Hood LE, Prusiner SB, Weissmann C: A cellular gene encodes scrapie PrP 27–30 protein. Cell 1985;40:735–746.
13   McKinley MP, Bolton DC, Prusiner SB: A protease-resistant protein is a structural component of the scrapie prion. Cell 1983;35:57–62.
14   Stahl N, Baldwin M, Teplow DB, Hood L, Gibson BW, Burlingame AL, Prusiner SB: Structural analysis of the scrapie prion protein using mass spectrometry and amino acid sequencing. Biochemistry 1983;32:1991–2002.
15   Prusiner SB: Molecular biology of prion diseases. Science 1991;252:1515–1522.
16   Prusiner SB: The prion diseases. Sci Am 1995;272:48–57.
17   Prusiner SB: in: Les Prix Nobel. Stockholm, The Nobel Foundation, pp 268–323.
18   Büeler H, Fischer M, Lang Y, Bluethmann H, Lipp HP, DeArmond SUJ, Prusiner SB, Aguet M, Weissmann C: Normal development and behaviour of mice lacking the neuronal cell surface PrP protein. Nature 1992;356:577–582.
19   Büehler H, Aguzzi A, Sailer A, Greiner RA, Autenried P, Aguet M, Weissmann C: Mice devoid of PrP are resistant to scrapie. Cell 1983;73:1339–1347.
20   Bruce ME, Dickinson AG: Biological evidence that the scrapie agent has an independent genome. J Gen Virol 1987;68:79–89.
21   Kimberlin RH, Cole S, Walker DA: Temporary and permanent modifications to a single strain of mouse scrapie in transmission to rats and hamsters. J Gen Virol 1987;68:1875–1881.
22   Riesner D: The search for a nucleic acid component to scrapie infectivity. Semin Virol 1991;2:215–226.
23   Meyer N, Rosenbaum V, Schmidt B, Gilles K, Mirenda C, Groth D, Prusiner SB, Riesner D: Search for a putative scrapie genome in purified prion fractions reveals a paucity of nucleic acids. J Gen Virol 1991;72:37–49.
24   Kellings K, Meyer N, Mirenda C, Prusiner SB, Riesner D: Further analysis of nucleic acids in purified scrapie prion preparations by improved return refocusing gel electrophoresis (RRGE). J Gen Virol 1992;73:1025–1029.

25   Kellings K, Prusiner SB, Riesner D: Nucleic acids in prion preparations: Unspecific background or essential component? Philos Trans R Soc Lond B 1994;343:425–430.

26   Riesner D, Kellings K, Meyer N, Mirenda C Prusiner SB: Nucleic acids and the scrapie agent; in Prusiner S, Collinge J, Powel J, Anderton B (eds): Prion Diseases in Humans and Animals. London, Horwood, 1992.

27   Hill AF, Collinge J: Strain variations and species barriers; in Rabenau HF, Cinatl J, Doerr HW (eds): Prions. A Challenge for Science, Medicine and Public Health System. Contrib Microbiol. Basel, Karger, 2001, vol 7, pp 48–57.

28   Knight R, Collins S: Human prion diseases: Cause, clinical and diagnostic aspects; in Rabenau HF, Cinatl J, Doerr HW (eds): Prions. A Challenge for Science, Medicine and Publich Health System. Contrib Microbiol. Basel, Karger, 2001, vol 7, pp 68–92.

29   Prusiner SB, Groth D, Serban A, Stahl N, Gabizon R: Attempts to restore scrapie prion infectivity after exposure to protein denaturants. Proc Natl Acad Sci USA 1993;90:793–2797.

30   Wille H, Zhang GF, Baldwin MA, Cohen FE Prusiner SB: Separation of scrapie prion infectivity from PrP amyloid polymers. J Mol Biol 1996;259:608–621.

31   Riesner D, Kellings K, Post K, Wille H, Serban H, Baldwin M, Groth D, Prusiner SB: Disruption of prion rods generates spherical particles composed of four to six PrP 27–30 molecules that have a high α-helical content and are non-infectious. J Virol 1996;70:1714–1722.

32   Riek R, Hornemann S, Wider G, Billeter M, Glockshuber R, Wüthrich K: NMR structure of the mouse prion protein domain PrP (121–231). Nature 1996;382:180–182.

33   James TL, Liu H, Ulyanov NB, Farr-Jones S, Zhang H, Donne D-G, Kaneko K, Groth D, Mehlhorn I, Prusiner SB, Cohen FE: Solution structure of a 142-residue recombinant prion protein corresponding to the infectious fragment of the scrapie isoform. Proc Natl Acad Sci USA 1997; 94:10086–10091.

34   Caughey BW, Dong A, Baht KS, Ernst D, Hayes SF, Caughey WS: Secondary structure analysis of the scrapie-associated protein PrP 27-30 in water by infrared spectroscopy. Biochemistry 1991; 30:7672–7680.

35   Pan KM, Baldwin M, Nguyen J, Gasset M, Serban A, Groth D, Mehlhorn I, Huang Z, Fletterick RJ, Cohen FE, Prusiner SB: Conversion of α-helices into β-sheets features in the formation of the scrapie prion proteins. Proc Natl Acad Sci USA 1983;90:10962–10966.

36   Kocisko DA, Come JH, Priola SA, Chesebro B, Raymond GJ, Lansbury PT Jr, Caughey B: Cell-free formation of protease-resistant prion protein. Nature 1994;370:471–474.

37   Hill AF, Antoniou M, Collinge J: Protease-resistant prion protein produced in vitro lacks detectable infectivity. J Gen Virol 1999;80:11–14.

38   Kaneko K, Peretz D, Pan KM, Blochberger TC, Wille H, Gabizon R, Griffith OH, Cohen FE, Baldwin MA, Prusiner SB: Prion protein (PrP) synthetic peptides induce cellular PrP to acquire properties of the scrapie isoform. Proc Natl Acad Sci USA 1995;91:11160–11164.

39   Post K, Pitschke M, Schäfer O, Wille H, Appel TR, Kirsch D, Mehlhorn I, Serban H, Prusiner SB, Riesner D: Rapid acquisition of β-sheet structure in the prion protein prior to multimer formation. Biol Chem 1998;379:1307–1317.

40   Post K, Brown DR, Groschup M, Kretzschmar H, Riesner D: Aggregated prion proteins containing β-sheet structures are sufficient to induce neurotoxic effects but are not infectious. Arch Virol, in press.

41   Klein TR, Kirsch D, Kaufmann R, Riesner D: Prion rods contain small amounts of two host sphingolipids as revealed by thin-layer chromatography and mass spectrometry. Biol Chem 1998; 379:655–666.

42   Appel TR, Dumpitak Ch, Mathiesen U, Riesner D: Prion rods contain an inert polysaccharide scaffold. Biol Chem 1999;380:1295–1306.

43   Prusiner SB, Scott M, Foster D, Pan KM, Groth D, Mirenda C, Torchia M, Yang SL, Serban D, Carlson GA, Hoppe PC, Westaway D, DeArmond SJ: Transgenetic studies implicate interactions between homologous PrP isoforms in scrapie prion replication. Cell 1990;63:673–686.

44   Come JH, Fraser PE, Lansbury PT: A kinetic model for amyloid formation in the prion diseases: Importance of seeding. Proc Natl Acad Sci USA 1993;90:5959–5963.

45 Telling GC, Scott M, Mastrianni J, Gabizon R, Torchia M, Cohen FE, DeArmond SJ, Prusiner SB: Prion propagation in mice expressing human and chimeric PrP transgenes implicates the interaction of cellular PrP with another protein. Cell 1995;83:79–90.

46 Huang Z, Prusiner SB, Cohen FE: Structures of prion proteins and conformational models for prion diseases; in Prusiner SB (ed): Prions Prions Prions. Berlin, Springer, 1996, pp 49–64.

47 Eigen M: Prionics or the kinetic basis of prion diseases. Biophys Chem 1996;63:A1–A18.

48 Prusiner SB: Inherited prion diseases. Proc Natl Acad Sci USA 1994;91:4611–4614.

Prof. Dr. Detlev Riesner, Institut für Physikalische Biologie,
Universitätsstrasse 1, D–40225 Düsseldorf (Germany)
Tel. +49 211 81 14840, Fax +49 211 81 15167, E-Mail riesner@biophys.uni-duesseldorf.de

Rabenau HF, Cinatl J, Doerr HW (eds): Prions. A Challenge for Science,
Medicine and Public Health System. Contrib Microb. Basel, Karger, 2001, vol 7, pp 21–31

[URE3] and [PSI]: Prions of *Saccharomyces cerevisiae*

Kimberly L. Taylor, Reed B. Wickner

Laboratory of Biochemistry and Genetics, National Institute of Diabetes,
Digestive and Kidney Diseases, National Institutes of Health, Bethesda, Md., USA

Prion, a term previously coined to describe the infectious material found in scrapie-associated filaments, is now more broadly used to describe infectious proteins. Prions act as hereditary material in that they have the ability to convert a normal, functional protein into an abnormal or prion form, which lacks the activity of the normal form [1]. Although the exact mechanism of prion propagation is not known, there is a growing body of evidence which suggests that protein misfolding or a conformational change is required to alter the normal form of a protein into its prion form. Two prion systems, [URE3] and [PSI] (table 1), have been identified in *Saccharomyces cerevisiae* and will be discussed in further detail.

Structure-Based Inheritance of Prions

[URE3] and [PSI] are heritable protein-based phenotypes apparently due to amyloid formation. Evidence from both Ure2p and Sup35p fibers suggests that amyloid formation is dictated by the self-association of the NH_2-terminal-prion-inducing region of these proteins, thus forming very stable backbones of the fibers. The COOH-terminal regions, which are not critical for the stability of the fibers, are then packed against the exterior of this core region. Although both the [URE3] and [PSI] systems show amyloid formation in vitro, it is not certain what the infectious or heritable material is in vivo. Amyloid formation may only be an endpoint whereas smaller fragments or precursors of these structures may act as seeds to carry out the transfer of heritable material from one protein to another.

Table 1. Comparison of the prion systems of *S. cerevisiae*

Prion system	[URE3]	[PSI]
Phenotype	nitrogen derepression	translation readthrough
Protein	Ure2p	Sup35p
Function of normal protein	nitrogen repression	translation termination
Modes of curing	growth on GuHCl expression of Ure2p-GFP fusions	growth on GuHCl osmotic shock overexpression of Hsp104 deletion of Hsp104
Maintenance gene	unknown	Hsp104
Prion inducing regions of proteins	NH$_2$-terminal residues 1–80 Asn rich	NH$_2$-terminal residues 1–123 Gln and Asn rich imperfect repeats of PQGGYQQYN
Filament forming regions of protein	NH$_2$-terminus	NH$_2$-terminus
Secondary structure of the prion domain	β-sheet	β-sheet
Protease resistance of prion protein	7–10 kDa NH$_2$-terminal fragment	transiently resistant NH$_2$-terminal region

This work shows that proteins can be genes, in analogy to the inheritance of cellular structure patterns shown by Beisson and Sonneborn [2]. As the yeast systems have provided some of the best evidence that there really are infectious proteins (prions), and that chaperones can play a critical role in prion phenomena, they will doubtless continue to be valuable in exploring the mechanisms of prion generation and propagation.

Discovery of the Non-Mendelian Genetic Element [URE3]

While studying uracil biosynthesis in *S. cerevisiae*, Lacroute [3] isolated mutants by selecting for the ability to take up ureidosuccinate (USA, product of the second step in the uracil pathway) on media containing ammonia, a readily metabolized nitrogen source. Most isolates were chromosomal, defining the *ure1* and *ure2* genes [4, 5], while some were dominant, showed irregular segregation in meiosis and were transmissible by cytoplasmic mixing (cytoduction), implying that this was caused by a nonchromosomal gene that he named [URE3] [3, 6]. The mutant *ure2* and [URE3] phenotypes were found to be the same.

The normal function of the *URE2* gene product, Ure2p, is to repress nitrogen catabolic genes under the control of transcriptional activator Gln3p, when a good nitrogen source is available [7]. Under conditions in which Ure2p is not functional, Gln3p activates the transcription of many genes involved in nitrogen catabolism. One of these genes *DAL5*, an allanoate permease, allows both allanoate and its structural homolog USA to be transported into the cell [8, 9]. Since USA is a biosynthetic precursor for uracil, USA must be directly supplied when this pathway is blocked by a *ura2* mutation in aspartate transcarbamylase. Thus *URE2* complementation and [URE3] are assayed in the laboratory by the ability of cells containing the *ura2* mutation to grow on media containing USA in the presence of a good nitrogen source.

[URE3] Fulfills the Genetic Criteria for Prions

Three genetic criteria that distinguish yeast prions from nucleic acid replicons were described by Wickner [1] in 1994 (fig. 1). To date, none of these criteria are satisfied by the mammalian transmissible spongiform encephalopathies (TSE). (1) Reversible curability. If a prion can be cured, then it should be able to arise again in the cured strain. [URE3] colonies that spontaneously arise on minimal media containing USA, can be cured by growth of cells on low concentration of guanidine HCl (M. Aigle, cited in [10]). [URE3] clones can be reisolated from the cured strain by growth on selective medium [1]. (2) Overproduction of the normal protein increases the chances for a prion to spontaneously arise. This has been shown for [URE3]; when the normal form of Ure2p is overproduced, the frequency of [URE3] is increased by 20–200-fold [1], and this effect is not due to the overproduction of the *URE2* mRNA or the presence of the gene in high copy number [11]. (3) The phenotype for the mutant gene and prion is the same, and the gene for the protein is necessary for the propagation of the prion. This is because a mutation in the gene and a prion change both result in a nonfunctional protein. In the case of [URE3], *ure2* mutants and [URE3] strains are both able to take up USA on minimal media containing a good nitrogen source, and *URE2* is necessary for the propagation of [URE3] [1, 6].

Further studies ruled out the possibility that [URE3] was a stable transcriptional state based on nitrogen regulation [11].

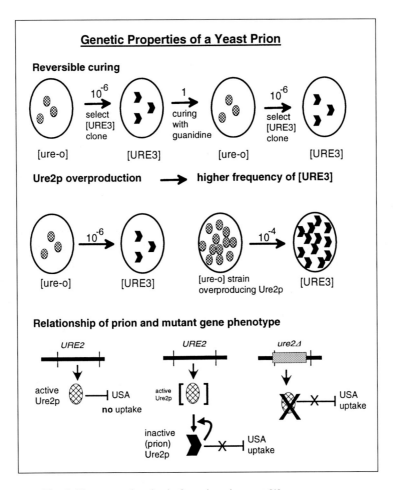

Fig. 1. Three genetic criteria for prions in yeast [1].

The Prion-Inducing and Nitrogen-Regulatory Domains of Urep

Genetic analysis shows that Ure2p is comprised of two domains, a prion-inducing domain and a nitrogen regulation domain [12, 13] (fig. 2). The NH_2-terminal region of Ure2p is rather unique in that it is composed of approximately 40% asparigine residues and to date shares no significant sequence homology with any other known or hypothetical proteins in the databases. Overexpression of the first 65 residues in the NH_2-terminal region increases the de novo formation of the [URE3] prion by over 1,000-fold the normal rate in the presence of a chromosomal copy of *URE2* [12] (fig. 2).

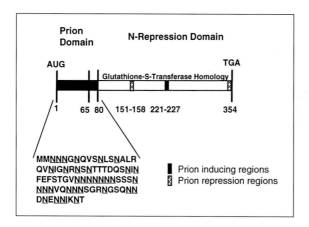

Fig. 2. Prion-inducing and nitrogen regulation domains of Ure2p. Genetic analysis of the *URE2* gene revealed regions of the protein involved in prion induction, prion repression and nitrogen regulation [12, 13].

This prion-inducing fragment fails to complement a *ure2Δ*, showing that it lacks the ability to repress nitrogen catabolism. Further analysis of Ure2p has shown that prion induction is enhanced when up to first 80 residues are overproduced [13]. Deletion of individual runs of asparagine within the prion domain reduce the efficiency of the prion-inducing activity [13].

Deletion of the first 65 residues of Ure2p leaves a COOH-terminal region, the nitrogen regulation domain, which is able to complement *ure2Δ* [12, 14], but is not able to induce the [URE3] prion in vivo. While the COOH-terminal region appears less efficient in carrying out its role in nitrogen regulation in the absence of the NH$_2$-terminal domain, the NH$_2$-terminal domain induces [URE3] more efficiently when the COOH-terminal domain is absent [12]. This suggests the two domains stabilize each other. In addition, two-hybrid analysis and affinity-binding experiments carried out by Fernandez-Bellot et al. [15] suggest that Ure2p interacts with itself through the NH$_2$-terminal domain (residues 1–63) and the COOH-terminal domain (152–354), confirming that the presence of both the prion-inducing domain and nitrogen regulation domain stabilize the normal form of Ure2p. Although the COOH-terminal region shares homology with the glutathione-S-transferases (GST [14]), to date no GST activity has been found [16]. Ure2p purified from *S. cerevisiae* forms dimers, suggesting that the COOH-terminal domain does not only function in nitrogen regulation, but may also function as a dimerization domain [17; Taylor and Wickner, unpubl. data].

Ure2p has two nonoverlapping regions which can promote [URE3] induction at a high efficiency [13]. When a fragment of Ure2p lacking the prion

Ure2p-GFP expressed in

| Guanidine-cured | [URE3] | [ure-o] |

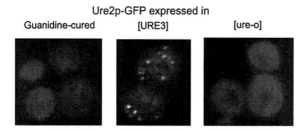

Fig. 3. Ure2p-GFP fusion proteins are aggregated in [URE3] cells. Aggregation of Ure2p was specifically detected in [URE3] strains when a Ure2p-GFP fusion was overexpressed from a single-copy plasmid with the URE2 promoter and was examined by fluorescence [18]. Wild-type [ure-o] and guanidine-cured strains show the Ure2p-GFP fusion evenly dispersed in the cytoplasm.

domain was further deleted at residues 151–158 and the seven COOH-terminal residues, the fragment was able to induce [URE3] at a high frequency. Deletion studies of residues 66–80 and the region near residue 224 revealed that these regions are necessary for this second prion-inducing activity.

URE2-GFP Fusions Aggregate in [URE3] Cells

Ure2p-green fluorescent protein (GFP) fusions expressed in *S. cerevisiae* were found to form intracellular aggregates in [URE3] cells (fig. 3) whereas in [ure-o] cells they were found to be dispersed throughout the cytoplasm [18]. Transformation of plasmids expressing fragments of Ure2p or Ure2-GFP fusion proteins into strains carrying the [URE3] phenotype often caused the strains to be cured. One plausible explanation for this phenomenon is that fragments of Ure2p or the fusion proteins may poison the crystal-like propagation of the prion form [18].

Prion Domain Initiates Amyloid Formation In Vitro Parallels [URE3]

When overexpressed, residues 1–65 of Ure2p, are sufficient to induce [URE3] formation in vivo [12]. In vitro the corresponding synthetic peptide, Ure2p$^{1-65}$, was shown to produce amyloid filaments 40–45 Å in diameter ([17], fig. 4a). Denatured Ure2p$^{1-65}$ when diluted into equal molar amounts of purified Ure2p was found to specifically precipitate full-length Ure2p to form thicker cofilaments 180–220 Å in width (fig. 4b). Without the added prion domain,

Prion domain = Ure2p$^{1-65}$ **Cofilaments = Ure2p$^{1-65}$+ Ure2p (full)**

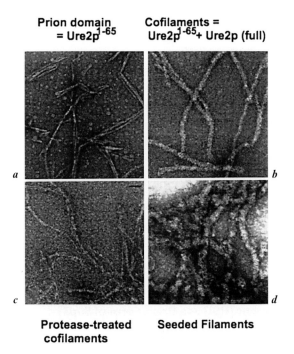

Protease-treated cofilaments   **Seeded Filaments**

Fig. 4. Ure2p forms amyloid in vitro [17]. *a* 45-ångström filaments formed by Ure2p$^{1-65}$. *b* Ure2p$^{1-65}$/Ure2p 1:1 cofilaments 200 Å in diameter. *c* Ure2p$^{1-65}$/Ure2p cofilaments digested with proteinase K leaving the NH$_2$-terminal prion domain in the form of a narrow filament. *d* 280–400 Å in diameter filaments of the native Ure2p formed by seeding with Ure2p$^{1-65}$/Ure2p cofilaments.

this native Ure2p purified from yeast was stable indefinitely, except for a nonspecific aggregation promoted by low salt or low pH conditions. Furthermore these cofilaments were able to seed filament formation by a solution of Ure2p forming even thicker fibers (280–400 Å) (fig. 4d). All of the filaments were found to bind the dye Congo Red, showing birefringence under polarized light [17].

Ure2p is more proteinase-K-resistant in extracts of [URE3] strains than wild-type strains [12]. Fragments of 30–32 kD transiently persist and give rise to smaller protease resistance fragments of approximately 7–10 kD which are quite stable. This pattern suggests that Ure2p has formed a specific structure in which part of the molecule is protected from protease, unlike nonspecific aggregates in which degradation is just generally slowed without accumulation of special stable fragments. Proteinase-K-treated Ure2p$^{1-65}$ filaments, Ure2p$^{1-65}$/Ure2p cofilaments and propagated Ure2p fibers show this same pattern of protease resistance, suggesting that the Ure2p incorporated into the filaments has adopted a similar conformation as the [URE3] form in cell-free extracts [17].

Amyloid is defined as a filamentous protein structure in β-sheet, resistant to proteases and showing yellow-green birefringence on staining with Congo red [19]. The Ure2p filaments described above have a high β-sheet content, are protease resistance as discussed, and show dramatic yellow-green birefringence when stained with Congo red [17], and thus have all the properties expected of amyloid.

The data support the notion that in vivo amyloid formation is responsible for [URE3]: (1) the in vivo prion domain of Ure2p is the promoter of amyloid formation in vitro, (2) the pattern of protease resistance of amyloid filaments resembles that seen in extracts of [URE3] strains, (3) Ure2p is specifically aggregated in [URE3] strains, and (4) the amyloid formation in vitro can propagate in a way analogous to the [URE3] in vivo [17].

Non-Mendelian Genetic Element [PSI] as a Prion Form of Sup35

Sup35p, a subunit of the translation release factor, recognizes termination codons and releases the completed peptides from the last tRNA [20, 21]. [PSI] is a non-Mendelian genetic element which, like *sup35* mutations, increases the strength of weak suppressor tRNAs [10, 22].

[PSI] satisfies the genetic criteria as a prion form of Sup35p [1]. [PSI] can be cured to [psi-] by growing in the presence of low concentration of guanidine hydrochloride or by high osmotic strength medium [23, 24]. But from these cured strains, can be isolated [PSI +] clones which have arisen de novo [25, 26]. The frequency of [PSI] increased by over 100-fold when Sup35p was overexpressed [26] and this is not caused by an excess of SUP35 DNA or mRNA [27], suggesting that Sup35p is necessary and sufficient for the propagation of the prion form. The phenotype of *sup35* mutants is similar to the phenotype of [PSI], and Sup35p is necessary for the propagation of [PSI] [28, 29].

Sup35 can be divided into three major regions, the NH_2-terminal, middle, and COOH-terminal regions. While the COOH terminal domain is essential for cell viability due to its function in translation termination, both the NH_2-terminus and middle region of Sup35p are not essential [29]. Genetic analysis of *SUP35* has shown that the prion-inducing domain of Sup35p is localized at the NH_2 terminus, a highly rich region of glutamine and asparagine containing several imperfect repeats of PQGGYQQYN [29, 30]. An allele of *SUP35*, PNM (Psi-No-More), alters the NH_2-terminal region of Sup35p, causing the loss of [PSI] in cells [28]. While cells with deletions in the NH_2-terminal region of Sup35p are no longer able to maintain [PSI] [29], transient overexpression of this domain increases the frequency of [PSI] heritable elements [30]. Thus the NH₂-terminal region of Sup35p is both necessary for maintenance and propagation of the prion form of Sup35p, and sufficient when overexpressed to make [PSI] appear.

This NH₂-terminal region of Sup35p also directs the aggregation of Sup35p specifically in [PSI+] strains and their extracts. First, Sup35p is found aggregated specifically in extracts of [PSI+] strains, and this requires the prion domain [31]. Second, fusions of the NH₂-terminal and middle regions of Sup35p with GFP were overexpressed in living yeast cells and were shown to form small aggregates in [PSI] cells but were shown to be evenly dispersed throughout the cells in [psi–] strains [32]. Third, the aggregation of native yeast Sup35p was found to propagate indefinitely in vitro in a reaction initiated by a small amount of [PSI+] extract [33].

Histidine-tagged NH₂-terminal and full-length Sup35p purified from *Echerichia coli* have been shown to form filaments in vitro [34, 35]. NH₂-terminal Sup35p filaments formed at pH 2.0 have the characteristics of amyloid, in that they are rich in β-sheet content, exhibit some protease resistance, and are able to bind the dye Congo Red, yielding apple-green birefringence under polarized light. Although filaments formed at pH 7.5 from the NH₂-terminal and middle portions of Sup35p are amyloid [S. Linquist, pers. commun.], fibers derived from full-length Sup35p as yet have not been reported to bind Congo Red with birefringence. Neither the middle nor the COOH-terminal regions of Sup35p form filaments on their own, thus, the NH₂-terminal region is necessary and sufficient for fiber formation in vitro.

Hsp104 Cures and Is Required for Propagation of [PSI]

Overexpression, as well as deletion of the chaperone Hsp104 eliminates [PSI] from cells [36, 37]. In the cured strains Sup35p becomes soluble and translation termination is restored. This state can be passed to its progeny even when the overexpression of Hsp 104 has ceased. The mechanisms by which normal amounts of HSP 104 allow propagation of [PSI] are not yet clear, but the data above showing filament formation from Sup35p in the absence of Hsp104 suggest that it does not have a direct role in the propagation reaction, but perhaps Hsp104 acts by breaking up the [PSI] aggregates of Sup35p ensuring segregation of aggregates to both daughter cells [38]. To date such a structural maintenance gene for [URE3] has not been discerned.

References

1   Wickner RB: Evidence for a prion analog in *S. cerevisiae*: The [URE3] non-Mendelian genetic element as an altered *URE2* protein. Science 1994;264:566–569.

2   Beisson J, Sonneborn TM: Cytoplasmic inheritance of the organization of the cell cortex in *Paramecium aurelia*. Proc Natl Acad Sci USA 1965;53:275–282.

3   Lacroute F: Non-Mendelian mutation allowing ureidosuccinic acid uptake in yeast. J Bacteriol 1971;106:519–522.

4   Drillien R, Aigle M, Lacroute F: Yeast mutants pleiotropically impaired in the regulation of the two glutamate dehydrogenases. Biochem Biophys Res Commun 1973;53:367–372.

5   Drillien R, Lacroute F: Ureidosuccinic acid uptake in yeast and some aspects of its regulation. J Bacteriol 1972;109:203–208.

6   Aigle M, Lacroute F: Genetical aspects of [URE3], a non-Mendelian, cytoplasmically inherited mutation in yeast. Molec Gen Genet 1975;136:327–335.

7   Courchesne WE, Magasanik B: Regulation of nitrogen assimilation in *Saccharomyces cerevisiae*: Roles of the *URE2* and *GLN3* genes. J Bacteriol 1988;170:708–713.

8   Rai R, Genbauffe F, Lea HZ, Cooper TG: Transcriptional regulation of the *DAL5* gene in *Saccharomyces cerevisiae*. J Bacteriol 1987;169:3521–3524.

9   Turoscy V, Cooper TG: Ureidosuccinate is transported by the allantoate transport system in *Saccharomyces cerevisiae*. J Bacteriol 1987;169:2598–2600.

10 Cox BS, Tuite MF, McLaughlin CS: The Psi factor of yeast: A problem in inheritance. Yeast 1988; 4:159–179.

11 Maison DC, Maddelein M-L, Wickner RB: The prion model for [URE3] of yeast: Spontaneous generation and requirements for propagation. Proc Natl Acad Sci USA 1997;94:12503–12508.

12 Masison DC, Wickner RB: Prion-inducing domain of yeast Ure2p and protease resistance of Ure2p in prion-containing cells. Science 1995;270:93–95.

13 Maddelein M-L, Wickner RB: Two prion-inducing regions of Ure2p are nonoverlapping. Mol Cell Biol 1999;19:4516–4524.

14 Coschigano PW, Magasanik B: The *URE2* gene product of *Saccharomyces cerevisiae* plays an important role in the cellular response to the nitrogen source and has homology to glutathione S-transferases. Moll Cell Biol 1991;11:822–832.

15 Fernandez-Bellot E, Guillemet E, Baudin-Baillieu A, Gaumer S, Komar AA, Cullin C: Characterization of the interaction domains of Ure2p, a prion-like protein of yeast. Biochem J 1999;338:403–407.

16 Choi JH, Lou W, Vancura A: A novel membrane-bound glutathione S-transferase functions in the stationary phase of the yeast *Saccharomyces cerevisiae*. J Biol Chem 1988;273:29915–29922.

17 Taylor KL, Cheng N, Williams RW, Steven AC, Wickner RB: Prion domain initiation of amyloid formation in vitro from native Ure2p. Science 1999;283:1339–1343.

18 Edskes HK, Gray VT, Wickner RB: The [URE3] prion is an aggregated form of Ure2p that can be cured by overexpression of Ure2p fragments. Proc Natl Acad Sci USA 1999;96:1498–1503.

19 Glenner GG: The bases of the staining of amyloid fibers: Their physico-chemical nature and the mechanism of their dye-substrate interaction. Prog Histochem Cytochem 1981;13:1–37.

20 Zhouravleva G, Frolova L, LeGoff X, LeGuellec R, Inge-Vectomov S, Kisselev L, Philippe M: Termination of translation in eukaryotes is governed by two interacting polypeptide chain release factors, eRF1 and eRF3. EMBO J 1995;14:4065–4072.

20 Stransfield I, Jones KM, Kushnirov VV, Dagkesamanskaya AR, Poznyakovski AI, Paushkin SV, Nierras CR, Cox BS, Ter-Avanesyan MD, Tuite MF: The products of the SUP45 (eRF1) and SUP35 genes interact to mediate translation termination in *Saccharomyces cerevisiae*. EMBO J 1995;14:4365–4373.

22 Cox BS: PSI, a cytoplasmic suppressor of super-suppressor in yeast. Heredity 1965;20:505–521.

23 Singh AC, Helms C, Sherman F: Mutation of the non-Mendelian suppressor [PSI] in yeast by hypertonic media. Proc Natl Acad Sci USA 1979;76:1952–1956.

24 Tuite MF, Mundy CR, Cox BS: Agents that cause a high frequency of genetic change from [*psi+*] to [*psi–*] in *Saccharomyces cerevisiae*. Genetics 1981;98:691–711.

25   Lund PM, Cox BS: Reversion analysis of [psi–] mutations in *Saccharomyces cerevisiae*. Genet Res 1981;37:173–182.

26   Chernoff YO, Derkach IL, Inge-Vechtomov SG: Multicopy SUP35 gene induces de novo appearance of psi-like factors in the yeast *Saccharomyces cerevisiae*. Curr Genet 1993;24:268–270.

27   Derkatch IL, Chernoff YO, Kushnirov VV, Inge-Vechtomov SG, Liebman SW: Genesis and variability of [PSI] prion factors in *Saccharomyces cerevisiae*. Genetics 1996;144:1375–1386.

28   Doel SM, McCready SJ, Nierras CR, Cox BS: The dominant PNM2–mutation which eliminates the [PSI] factor of *Saccharomyces cerevisiae* is the result of a missense mutation in the *SUP35* gene. Genetics 1994;137:659–670.

29   Ter-Avanesyan A, Dagkesamanskaya AR, Kushnirov VV, Smirnov VN: The *SUP35* omnipotent suppressor gene is involved in the maintenance of the non-Mendelian determinant [psi+] in the yeast *Saccharomyces cerevisiae*. Genetics 1994;137:671–676.

30   Derkatch IL, Bradley ME, Zhou P, Chernoff YO, Liebman SW: Genetic and environmental factors affecting the de novo appearance of the [PSI+] prion in *Saccharomyces cerevisiae*. Genetics 1997; 147:507–519.

31   Paushkin SV, Kushnirov VV, Smirnov VN, Ter-Avanesyan MD: Propagation of the yeast prion-like [psi+] determinant is mediated by oligomerization of the *SUP35*-encoded polypeptide chain release factor. EMBO J 1996;15:3127–3134.

32   Patino MM, Liu J-J, Glover JR, Lindquist S: Support for the prion hypothesis for inheritance of phenotypic trait in yeast. Science 1996;273:622–626.

33   Paushkin SV, Kushnirov VV, Smirnov VN, Ter-Avanesyan MD: Interaction between yeast Sup45p (eRF1) and Sup35p (eRF3) polypeptide chain release factors: Implications for prion-dependent regulation. Mol Cell Biol 1997;17:2798–2805.

34   King C-Y, Tittmann P, Gross H, Gebert R, Aebi M, Wuthrich K: Prion-inducing domain 2-1114 of yeast Sup35 protein transforms in vitro into amyloid-like filaments. Proc Natl Acad Sci USA 1997;94:6618–6622.

35   Glover JR, Kowal AS, Shirmer EC, Patino MM, Liu J-J, Lindquist S: Self-seeded fibers formed by Sup35, the protein determinant of [PSI+], a heritable prion-like factor of *S. cerevisiae*. Cell 1997;89:811–819.

36   Chernoff YO, Ono BI: Dosage-dependent modifiers of PSI-dependent omnipotent suppression in yeast; in Brown AJP, Tuite MF, McCarthy JEG (eds): Protein Synthesis and Targeting in Yeast. Berlin, Springer, 1992, pp 101–107.

37   Chernoff YO, Lindquist SL, Ono B-I, Inge-Vechtomov SG, Liebman SW: Role of the chaperone protein Hsp104 in propagation of the yeast prion-like factor [psi+]. Science 1995;268:880–884.

38   Paushkin SV, Kushnirov VV, Smirnov VN, Ter-Avanesyan MD: In vitro propagation of the prion-like state of yeast Sup35 protein. Science 1997;277:381–383.

Dr. Reed B. Wickner, Bldg. 8, Room 225, N.I.H.
8 Center Drive MSC 0830 Bethesda, MD 20892–0830 (USA)
Tel. (301) 496 3452, Fax (301) 402 0240, E-Mail wickner@helix.nih.gov

Rabenau HF, Cinatl J, Doerr HW (eds): Prions. A Challenge for Science,
Medicine and Public Health System. Contrib Microb. Basel, Karger, 2001, vol 7, pp 32–47

····················

Structural Biology of Prions

Roberto Cappai [a], *Michael F. Jobling* [a, b], *Colin J. Barrow* [b],
Steven Collins [a]

[a] Department of Pathology, The University of Melbourne, and Mental Health
Research Institute, and
[b] School of Chemistry, The University of Melbourne, Parkville, Victoria, Australia

The transmissible spongiform encephalopathies (TSEs) constitute a group
of fatal neurodegenerative diseases occurring in humans – Creutzfeldt-Jakob
disease (CJD), Gerstmann-Sträussler-Scheinker syndrome (GSS), variant
CJD, kuru and fatal familial insomnia – and animals – e.g. bovine spongiform
encephalopathy (BSE) and scrapie. A considerable body of data supports the
model that the TSE infectious agent known as the prion, is a self-replicating
protein devoid of nucleic acid [1, 2]. According to the protein-only hypothesis,
prion propagation involves changing the conformation of the normal cellular
prion protein (PrP^C) into an infectious isoform (PrP^{TSE} or also called PrP^{Sc}).
The two conformers have distinct biochemical and biophysical properties but
share the same primary sequence. While PrP^C is relatively detergent soluble
and sensitive to proteinase K digestion, PrP^{TSE} aggregates into rods or filaments
and is detergent insoluble and resistant to proteinase K digestion [3]. These
biochemical differences are believed to reflect the different secondary and
tertiary structures of PrP^C and PrP^{TSE} [4]. The majority of human TSE cases
($>85\%$) are sporadic with an unknown aetiology, and approximately 12–14%
are caused by mutations in the PrP gene (PRNP). These mutations may
promote prion formation by destabilizing PrP^C and facilitating the transition
to PrP^{TSE}. The presence of different prion strains which manifest themselves
as distinct clinico-pathological phenotypes has been difficult to accommodate
within the protein-only hypothesis, but it may be related to different PrP^{TSE}
conformers. If validated, this suggests PrP^C possesses considerable structural
plasticity. Therefore, a detailed description of the structural properties and
the cellular and molecular mechanisms of interconversion of PrP^C to PrP^{TSE}
is central to understanding prion disease biology. Considerable advances have

been made towards achieving this goal with the determination of the three-dimensional structure of the globular domain of PrP^C. However, major gaps still remain, with the most salient being the structure of PrP^{TSE}. This has been hampered by the insolubility of PrP^{TSE} which prevents it from being amenable to the same analysis as PrP^C.

The Primary Structure of the Prion Protein

There are three commonly described PrP species, the normal non-infectious cellular isoform PrP^C, the full length infectious form PrP^{TSE} and PrP27–30 which is obtained after partial proteolysis of PrP^{TSE}. The PrP27–30 protein is an N-terminally truncated molecule that commences at residue 90 and represents the infectious protease resistant core of PrP^{TSE}. The purification of PrP27–30 from infected hamster brain enabled the amino terminus and internal peptide fragments to be sequenced [5]. Degenerate oligonucleotides based on these sequences were used to screen an infected hamster brain cDNA library, and a PRNP cDNA clone was isolated [5]. Southern blot analysis of genomic DNA indicated that the PrP sequence was present as a single-copy gene. A comparison of genomic DNA from normal and scrapie infected hamster brain gave an identical restriction pattern, indicating that infectivity did not involve chromosomal rearrangements. The human PRNP gene is localized to the short arm of chromosome 20 [6]. The PrP open reading frame is encompassed on a single exon of approximately 2 kb that is spliced onto two small exons located 10 kb upstream [7, 8]. Expression occurs in most adult tissues, except the liver, with the highest levels being in the brain. The promoter sequence of hamster PRNP resembles that of a housekeeping gene being GC rich and lacking an apparent TATA box [7]. Multiple transcriptional start sites were identified 140–160 bp upstream of the ATG start codon. There are no differences in PRNP mRNA expression levels between infected and uninfected brains, or during infection [5, 9].

The nascent PrP molecule is approximately 250 amino acids in length. It has an amino-terminal signal peptide and a hydrophobic carboxy-terminal domain for membrane attachment via a glycosylphosphatidylinositol anchor (fig. 1). There are four copies of an octapeptide repeat PHGGGWGQ (codons 60–91, numbering based on human sequence) and an analogous nonapeptide PQGGGGWGQ (codons 51–59) in the N-terminal region. There is a single disulphide bond between cysteines 179 and 214 and two N-glycosylation sites at asparagines 182 and 198. A hydrophobic sequence (codons 113–135) in the middle of the protein may exist as a transmembranous region in some PrP isoforms [10]. Transgenic mice expressing PRNP with mutations which favour

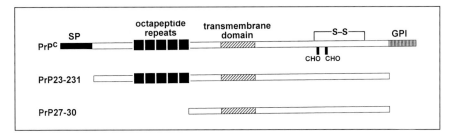

Fig. 1. Schematic representation of the main primary structural features of PrP and PrP27–30. SP represents the signal peptide; S–S indicates the disulphide bond; CHO represents the N-linked glycosylation sites; GPI represents the glycosylphosphatidylinositol anchor.

the transmembrane form resulted in clinico-pathological features that are found in some inherited human prion diseases. To understand the evolutionary conservation of PRNP, the gene was sequenced from a diverse range of species including chickens, rodents, sheep, primates and humans [11]. There is high amino acid sequence identity within mammals and birds, but low homology between birds and mammals. However, overall there was strong conservation of key structural elements between mammals and birds. The N-terminal region tends to differ by insertions and deletions in the octapeptide repeat domain. In contrast, the C-terminal region (91–231) varies by point mutations. Interestingly, there is high conservation of sequences (residues 23–90) which are not associated with either the structural elements or the infectious portion of the molecule.

Ablation of the PRNP gene by homologous recombination resulted in PrP knockout mice (PrP$^{o/o}$) that were viable and overall phenotypically normal [12]. The PrP$^{o/o}$ mice are completely resistant to infection by PrP^{TSE}. Another line of PrP$^{o/o}$ mice was generated that developed late-onset ataxia [13]. The reason for the ataxia phenotype may be explained by the presence of a PRNP homologue located 16 kb downstream of the PrP gene [14]. The gene is designated *Prnd* and the protein product doppel (downstream PrP like). It encodes a 179-residue protein with 25% amino acid identity to PrP. *Prnd* gene expression is upregulated in PrP$^{o/o}$ mice with late-onset ataxia. This is probably due to the deletion of the PRNP exon 3 splice acceptor site in these mice and this may alter transcriptional regulatory elements of *Prnd*. It remains to be determined if doppel has a modulating effect on the clinical phenotype in prion disease.

While the precise function of PrP^C is unknown, a number of activities have been associated with PrP^C including a role in maintaining cell survival

[15, 16], promoting neurite outgrowth [17] and modulating synaptic transmission [18]. Two regions of PrP have been studied for their toxic and metal-binding properties. A peptide fragment encompassing residues 106–126 of human PrP has been shown to be highly fibrillogenic and toxic to neurons in vitro [19, 20]. Since PrP106–126 toxicity mimics PrP[TSE] by requiring the expression of PrP[C] to cause cell death [21], this peptide represents a suitable model to study the toxic properties of PrP[TSE]. This PrP106–126 sequence contains the amyloidogenic palindrome AGAAAAGA from position 113 to 120 [19]. Based on the NMR structure, this sequence is part of, or adjacent to, the first of the two very short β-sheets in PrP[C] [22–26]. This region of PrP[C] is highly hydrophobic and has been shown to exhibit considerable flexibility and structural plasticity [23]. The structure of the PrP106–126 peptide is modulated by pH and has more β-sheet at pH 5 than at pH 7 [27–29]. Additionally, in the presence of lipids (at pH 7.4) it acquires a predominantly β-sheet conformation [28]. The hydrophobic AGAAAGA palindromic sequence is a key modulator of PrP106–126 toxicity and amyloidogenicity [29]. Hydrophilic substitutions in this hydrophobic core abolished neurotoxicity and this effect correlated with decreases in β-sheet content, aggregability and fibrillogenesis. The hydrophobic C-terminal valines, and the palindrome region from Ala113 to Ala120 of PrP106–126 are integral to the folding and/or stabilization of a β-sheeted aggregate. The behaviour of PrP106–126 is similar to Alzheimer's disease amyloid beta (Aβ) peptide and supports the view of a common structure-function mechanism of amyloid generation in spongiform encephalopathies and Alzheimer's disease [30–32].

Expansions in the octapeptide repeat number are linked to familial cases of prion disease. Interestingly, deletions of octapeptide repeats is not associated with disease [33]. While this iterative sequence has been shown to bind copper [34–37] there is some controversy towards the stoichiometry of copper binding to PrP[C] with values ranging from 1.8 to 5.6 [36, 38]. Using synthetic peptides with different numbers of octarepeats, it was established that a two-octarepeat peptide binds one copper ion while three- and four-octarepeat peptides cooperatively bound three and four copper ions, respectively. All the peptides bound copper with a K_d of 6 μM [39]. The interaction between the peptides and copper was pH dependent, with no binding occurring below pH 6. The apo form of the peptide was shown to be unstructured [37, 39], but the effect of copper binding on the structure of the octarepeat peptide is unclear. One study showed that the addition of copper results in the formation of turns or structured loops [39]. In contrast, another study found that the addition of copper promoted α-helix formation [37]. Such differences may reflect buffer-dependent effects as similar studies using a Tris buffer, as opposed to phosphate buffers, did not alter octarepeat structure [35]. It was proposed that Tris-based buffers may be competing for copper at pH 7.4 [39]. While the analysis of copper binding to full-length PrP^C is problematic, due to its propensity to aggregate in the presence of copper, it has been possible to show that full-length PrP has a $K_d = 14$ μM [36].

Two different models for the PrP octapeptide-Cu complex have been proposed. One model suggests the binding occurs via the N_π atom from the imidazole side chain in histidine together with two deprotonated main-chain amide nitrogens from the triglycines [40]. At pH 6, the copper-amide linkages are broken and the copper becomes coordinated by an interpeptide linkage via the histidines. It was suggested that the changes in pH of brain micro-environments, as occurs from the cell surface to the endosome, could influence PrP-Cu binding and therefore its structure [40]. The second model of the octarepeat-Cu complex was based on nuclear magnetic resonance (NMR) and electron spin resonance studies that showed a tetragonal coordination indicative of a three nitrogen:one oxygen coordination site [39]. A four-octarepeat peptide would coordinate the four copper ions with nitrogens from the histidine imidazole group from two adjacent repeats and a nitrogen from the amide backbone. The final ligand, an oxygen molecule, would be from the buffer, most likely water. A two-octarepeat peptide would involve a similar arrangement [39].

Acknowledging that the data strongly support PrP being a cuproprotein, it remains to be determined whether native brain-derived PrP contains bound copper or other metals. This is particularly important since it has been shown that manganese competes with copper for binding to PrP in vitro [41]. In vivo studies in PrP knockout mice showed a decrease in synaptic copper levels [38]. However, these results have been disputed by another group who failed to detect a difference between PrP knockout and wild-type mice [42]. Supporting the potential importance of metals in prion biology is the finding that treating brain-derived PrP^{TSE} with chelators promoted the switching between strain types [43]. This suggests metals may play a role in prion biogenesis by modulating the conformation of PrP^{TSE} and/or PrP^C and supports the view that metals may modulate the property of proteins that are central to a number of neuro-degenerative diseases [44].

Secondary Structure Analysis of the Prion Protein

The purification of PrP^C and PrP^{TSE} under non-denaturing conditions allowed their secondary structures to be studied [4]. Fourier-transformed infrared spectroscopy (FTIR) and circular dichroism (CD) showed that PrP^C had a high α-helical content ($\sim 42\%$) and was almost devoid of β-sheet (3%).

In contrast, PrP^{TSE} exhibited a high amount of β-sheet (43%) and decreased levels of α-helix (32%). The incorporation of PrP^C or PrP27–30 into liposomes caused major alterations to their protein structure. Liposome encapsulated PrP^C had 34% β-sheet, 20% α-helix and the remaining 46% was both random coil and β-turns [45]. In contrast, liposome incorporated PrP27–30 had 43% β-sheet, 57% β-turns and random coil, and no α-helix. Although the transition of PrP^C to PrP^{TSE} appears to involve major increases in β-sheet structure and reductions in α-helix [4], the same data indicate that this conformational change involves regions other than the α-helices as a considerable level of α-helix remains in PrP^{TSE}. Therefore, PrP^C to PrP^{TSE} biogenesis must also involve the conversion of the coils and random regions to β-sheet. Alternatively, the α-helices in PrP^C may be converted to β-sheet concomitant with other less ordered regions of PrP^C converting to α-helix.

Analysis of PrP^{TSE} and PrP27–30 folding intermediates generated in guanidinium hydrochloride (Gdn-HCl) showed that PrP^{TSE} and PrP27–30 dissociate into monomers via a cooperative two-step transition from aggregate to monomer [45, 46]. In contrast to PrP^{TSE}, PrP27–30 appears to unfold via a stable intermediate(s) with a greater thermodynamic stability. Therefore, the amino-terminal region affects the global secondary structure and behaviour of the carboxy-terminal segment. The dissociation and unfolding of PrP^{TSE} may occur via a sequential four-step pathway of aggregates involving dissociation to folded monomers, partially unfolded intermediates and finally an unfolded monomer [45]. The PrP27–30 intermediate species is a compact, metastable hydrophobic molecule with a significant proportion of non-β-sheet secondary structure and little tertiary structure interaction [46]. It was proposed that this structure has the characteristics of an aggregated molten globule folding intermediate [46]. The role and possible mechanism by which this intermediate is generated in vivo is not known, but its induction by low pH in the presence of salts implicates particular organelles or general cellular acidosis as possible factors.

Nuclear Magnetic Resonance Structure of the Normal Prion Protein

The determination of the three-dimensional structure of PrP^C was made possible by improvements in the expression of recombinant PrP in sufficient quantities to permit NMR studies [47, 48] (fig. 2). The first structure to be solved, due to its solubility and high-level expression in bacteria, corresponded to mouse PrP (MoPrP) residues 123–231 [22]. This C-terminal fragment contained three α-helices and two short β-strands (128–131 and 161–164). The α-helices (helix-2, 179–193 and helix-3 at 200–217) corresponded to those

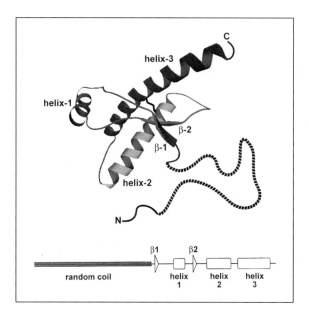

Fig. 2. Three-dimensional NMR structure of human PrP23–230 [26]. The three α-helices (helix-1, helix-2 and helix-3) and the two β-sheets (β-1 and β-2) are labelled. The disulphide bond joining helix-2 and helix-3 is represented by the black angular line. The N-terminus from 23 to 124 is unstructured and contains the octapeptide repeats (residues 51–91) which are represented by the broken line. A schematic of the secondary structure elements is shown below. The random coil is residues 23–124. β1 is 128–131 and β2 is 161–164. Helix-1, helix-2 and helix-3 are residues 144–154, 173–194 and 200–228, respectively.

predicted by the molecular-modelling approach [49]. However, helix-1 (144–154) and the two β-strands were not previously predicted. The second and third helices are linked by a disulphide bond to form a V-shaped pseudo-cyclic structure upon which the β-sheet and the first helix are anchored. The surface contains some hydrophobic patches near the β-sheet and the loop preceding helix-1 and an uneven distribution of positively and negatively charged residues. The invariant residues among the mammalian species were located in the hydrophobic core and presumably stabilize the overall fold of the protein. The familial point mutations are located within, or lie adjacent to the β-sheet and the second and third α-helical regions. This has obvious implications for their postulated effect on promoting PrP^C to PrP^{TSE} conversion by destabilizing these structural elements.

After this initial NMR report, the structure of PrP^C from other species and of longer lengths were solved, including full length $MoPrP^C$ (PrP23–231) [25], Syrian hamster (SHaPrP) SHaPrP29–231 [23], SHaPrP90–231 [24, 50], and human PrP (HuPrP) HuPrP23–230, HuPrP121–230 and HuPrP90–230 [26]. There was good overall consensus amongst the different structures with three α-helical and two short β-sheet regions. The N-terminal region (codons 21–120), which contains the octapeptide repeats, was shown to behave as a highly flexible tail with no apparent secondary structure. In SHaPrP, residues 113–128 formed a hydrophobic cluster with many side chain interactions [50] and this was adjacent to an irregular β-sheet region. Minor differences existed between the full-length and C-terminal fragments of SHaPrP and MoPrP, suggesting that there may be transient interactions between the structured core and the unstructured N-terminal domain [23]. Further refinement of the SHaPrP90–231 structure [50] extended helix-3 by 10 residues (200–227) and demonstrated many medium- and long-range nuclear Overhauser effects within the 113–128 hydrophobic cluster, indicating a definite structural inclination within this region. The presence of the N-terminal flexible tail stabilizes helix-3 and increases the length of this helix. The main difference between HuPrP, MoPrP and SHaPrP was in the backbone conformations of helix-3 and the loop joining the second β-strand and helix-2 [26]. HuPrP had features in common with both SHaPrP and MoPrP. The HuPrP and SHaPrP structures shared a longer helix-3, while the disordered loop in HuPrP from 167–171 is similar to that observed for MoPrP. The degree of disorder of the helix-2 and helix-3 residues is regulated by the length of the N-terminal tail. These species-based structural variations are located on the surface of PrP^C and may be responsible for modulating transmission across the species barrier.

The PrP sequence from 23 different mammalian species was modelled onto the MoPrP121–231 NMR structure [51]. It was found that the 103 residue segment from 124 to 226 was highly conserved with substitutions at only 18 residues. These substitutions localized to three distinct regions termed A, B and C. The A region is the loop between the second β-strand and the helix-2 and includes a small portion of the C-terminal tail. Seven out of nine substitutions within or near the A region involved changes in either side chain polarity or changes leading to significant electrostatic modification. Since this region localizes to the PrP surface, it may be involved in protein-protein interactions. Therefore, the sequence differences may alter intermolecular recognition by changing surface hydrogen bonding patterns. The B region involves five hydrophobic residues located where helix-1 interfaces with the rest of the protein. This region has limited surface exposure and is not a typical binding site, but it is proposed that slight alterations in conformation may expose these hydrophobic residues and allow them to promote PrP self association. The C region is located at the two ends of helix-1 and also represents a potential protein-protein binding site. The variant amino acids in the C region could modify the specificity of the intermolecular interactions. The regions which lacked substitutions include the two β-sheets, the loop joining helices-2 and -3, the C-terminal two turns of helix-2 and the central two turns of helix-1. Conservation of this region is presumably necessary to maintain the overall local three-dimensional structure and to facilitate intermolecular interactions to either itself or its ligand(s).

The Influence of Familial Mutations on the Structure of the Prion Protein

The pathophysiological effects of familial TSE mutations have been studied in both transgenic mice and cell culture models. Transgenic mice expressing MoPrP with the GSS proline to leucine mutation at residue 102 (residue 101 in mice) undergo spontaneous vacuolar neurodegeneration [52]. Furthermore, brain extracts from affected mice could transmit disease to recipient mice expressing the MoPrP-P101L mutation suggesting the generation of de novo prion disease [53]. However, the successful modelling of other PRNP mutations in transgenic mice has not been reported. The expression of MoPrP with an expanded octapeptide repeat resulted in PrP with altered biochemical properties but no transmission data were reported [54].

In cell culture, the expression of MoPrP harbouring the FFI-associated D178N mutation in Chinese hamster ovary (CHO) cells resulted in proteinase-K-resistant and detergent-insoluble PrP [55]. In contrast, expression of HuPrP-D178N in the human neuronal M-17 cell line resulted in proteinase-K-sensitive PrP [56]. This suggests differences in cell types and/or PrP species sequence can result in different biochemical properties. As already mentioned, based on the NMR structure of full-length or large fragments of mouse and hamster PrP [22–25], a number of the mutations are located close to the α-helical and β-sheet regions, suggesting they may be critically positioned to facilitate structural changes or alter the ligand-binding properties of PrP. Biophysical studies of recombinant proteins containing TSE mutations have shown effects on PrP structure and aggregation [57, 58]. Bacterially expressed MoPrP23–231-P101L was shown to alter the relative stability of the mutant protein as compared to the wild-type MoPrP23–231 protein [58]. The P101L mutation lowered the amount of α-helix and upon thermal denaturation the MoPrP23–231-P101L protein was more resistant to unravelling to random coil. Furthermore, wild-type MoPrP23–231 unfolded in a single-step process from helical to random coil as shown by the presence of an isodichroic point. In contrast, the MoPrP23–231-P101L protein did not have an isodichroic point, indicating that a metastable intermediate is likely to exist along the PrP-P101L unfolding pathway.

The structural stability of recombinant HuPrP27–30 (residues 90–231) encoding either the P102L mutation, the D178N mutation, or the CJD-associated E200K mutation, was measured under different Gdn-HCl concentrations and pHs [57]. At neutral pH, the E200K mutation caused a small reduction in the thermodynamic stability of the protein, but similar changes were not observed in the P102L and D178N proteins. At acidic pH, the PrP molecules unfolded via a stable intermediate and there was no observable difference between mutant and wild-type PrP. This is different to the behaviour of SHaPrP90–231, which showed irreversibility of thermal denaturation [59]. It was suggested that the pathogenic mutations may not affect the thermodynamic stability of PrP^C and cannot be used to rationalize the destabilization of PrP^C and its conversion into PrP^{TSE} [57]. This highlights the variation that can occur when comparing different species and fragments of PrP, prompting caution when attempting generalized conclusions.

Deciphering the Structure of Infectious Prion Protein with Antibodies

PrP-specific antibodies are becoming increasingly useful tools in PrP structural studies. The partial denaturation of PrP^{TSE} in Gdn-HCl increases its sensitivity to protease digestion and results in cleavage at codon 115 (as opposed to codon 90 without Gdn-HCl) and some proteolysis of the C-terminal residues 217–232 [60]. The partial denaturation of PrP^{TSE} was reversible since it could be refolded to a species which was only digested to residue 90 following dilution of the denaturant. The PrP27–30 protein displayed similar unfolding-refolding properties to PrP^{TSE} indicating that the N-terminal 89 residues are not necessary for the refolding reaction. The PrP sequence from 115 to 217–232 was the most resistant to unfolding, and the unfolding of this region beyond a threshold amount of denaturant prevented refolding. Recombinant antibody fragments (Fabs) raised against liposome incorporated PrP27–30, reacted with epitopes encompassed by residues 90–120 [61]. These epitopes are largely inaccessible in PrP27–30 but were exposed following denaturation with 3 M Gdn-thiocyanate. In contrast C-terminal epitopes are accessible in both PrP^C and PrP^{TSE}. This indicates that the N-terminal residues 90–120 are intimately involved in PrP^C to PrP^{TSE} conversion. This is supported by data showing that deleting any part of the 90–120 sequence perturbs PrP^{TSE} formation [62]. The reactivity of partially denatured PrP^{TSE} to the 3F4 monoclonal antibody formed the basis of a conformation-dependent immunoassay that could distinguish eight different prion strains propagated in Syrian hamsters [63]. This supports the model that strain types can correlate with different conformations of PrP molecules. If validated, this would confirm PrP as a molecule capable of assuming a number of different stable conformations.

Prion Protein Folding Pathways and Their Conformational Properties

The mobility of PrP^C as measured by hydrogen/deuterium exchange showed that in the unfolded state only a 10-residue region surrounding the disulphide bond retained its native structure [64]. There was no evidence of a highly organized folding intermediate. This suggests the conversion of PrP^C to PrP^{TSE} proceeds via a highly unfolded state which retains only a 10-residue nucleus rather than via a predominately organized folding intermediate which retains the α-helical regions. PrP conformation is influenced by pH as an acidic environment induces a dramatic increase in the exposure of hydrophobic regions on the surface of the protein [65]. At pH 5 to 7, HuPrP90–231 unfolds by a two-state pathway, but at lower pH, a three-state unfolding is observed with a stable transition intermediate [65]. This intermediate is rich in β-sheet structure and maintains a high degree of solubility, which may be the precursor for the initiation of PrP^{TSE} aggregates. While it was suggested that PrP is more likely to encounter lower pHs in a cellular environment [65], it was also proposed that the aqueous buffers used in biophysical studies are not physiologically relevant [66]. This was explored by measuring the interaction between PrP and lipid membranes and it was found that the membrane-bound form had a structurally less stable C-terminal domain and the flexible N-terminal became ordered [66]. This would be consistent with the N-terminal residues, 90–120, being necessary for PrP^{TSE} formation [61].

The folding kinetics of MoPrP121–231, as measured by tryptophan fluorescence, indicated that this fragment unfolded and refolded in the submillisecond scale at 22 °C [67] which is extremely rapid for a protein containing a disulphide bond. In contrast, the kinetics of unfolding and refolding measured at 4 °C showed that MoPrP121–231 folds with a half-life of 170 µs, with no obvious intermediates detected. It was concluded that the conversion of PrP^C to PrP^{TSE} does not involve a kinetic folding intermediate, but rather a thermodynamic folding intermediate. A possible folding intermediate was identified by reducing the disulphide bond of recombinant HuPrP91–231 in a low ionic strength buffer at acid pH. These conditions generated a highly soluble monomeric species which was rich in β-structure and partially proteinase K resistant [68]. This species, termed β-PrP, contained two unstructured regions (91–126 and 229–230) which were also unstructured in PrP^C [22] and suggests the rearrangement of PrP^C to β-PrP occurs in the structured regions of PrP^C. Increasing the pH to 8.0 caused the slow (over a number of days) conversion of β-PrP back to PrP^C. Raising the ionic strength to physiological concentrations resulted in β-PrP to form fibrils with scrapie amyloid-like morphology and dimensions. The buffer conditions for β-PrP formation are similar to those of endosomal organelles, which suggests this may be a site of PrP^{TSE} propagation.

Tertiary Structure of the Infectious Prion Protein

To date, the three-dimensional structure of PrP^{TSE} has not been reported. Secondary structure measurements of PrP^{TSE} have identified a high amount of β-structure compared with PrP^C which is predominately α-helix [4]. PrP27–30 contains even higher levels of β-structure correlating with its amyloidogenic propensity and subsequent increase in intermolecular hydrogen bonding. X-ray diffraction studies on hamster PrP27–30 shows β-sheet conformation, consistent with those observed by CD and FTIR [69]. The intersheet spacing of PrP27–30 is 8–9 Å which is small in comparison to other amyloids such the Aβ peptide of Alzheimer's disease. This suggests the PrP27–30 prion rods are packed via small side chains. Residues 113–127 of PrP are devoid of bulky aliphatic and aromatic residues, and this region could form the majority of the amyloidogenic core. It was established that PrP27–30, following solubilization with different solvents and exposure to reverse micelles, could form two- and three-dimensional microcrystals in the presence of uranyl salts and in acidic pH [70]. Electron diffraction data indicated at least four different types of crystal lattices were present. Although the suitability of these crystals for structural determinations remains to be clarified, it is an encouraging step toward solving the structure of PrP^{TSE}.

Conclusion

If the protein-only prion hypothesis proves correct, then the above-mentioned studies will be invaluable not only for understanding the molecular processes which underlie the SEs but also for understanding protein folding in general. The solving of the solution structure of PrP^C provides one part of the equation, but our lack of knowledge of the three-dimensional structure of PrP^{TSE} remains a clear deficiency. Deciphering the structure of PrP^{TSE} is vital as it will identify which regions of PrP^C are altered and therefore lead to a better understanding of the mechanism of PrP^C to PrP^{TSE} conversion. It will also help to explain the molecular basis of phenotypic strain variation and the nature of the 'toxic species'. Ultimately, a structure of PrP^{TSE} may permit the design of therapeutic agents that favourably perturb the transformation reaction and hence disease.

Acknowledgment

We thank Dr. Andrew Hill for critically reading the manuscript.

References

1   Prusiner SB: Novel proteinaceous infectious particles cause scrapie. Science 1982;216:136–144.
2   Prusiner SB: Prions. Proc Natl Acad Sci USA 1998;95:13363–13383.
3   Meyer RK, McKinley MP, Bowman KA, Braunfeld MB, Barry RA, Prusiner SB: Separation and properties of cellular and scrapie prion proteins. Proc Natl Acad Sci USA 1986;83:2310–2314.
4   Pan KM, Baldwin M, Nguyen J, Gasset M, Serban A, Groth D, Mehlhorn I, Huang Z, Fletterick RJ, Cohen FE, Prusiner SB: Conversion of α-helices into β-sheets features in the formation of the scrapie prion proteins. Proc Natl Acad Sci USA 1993;90:10962–10966.
5   Oesch B, Westaway D, Wälchli M, McKinley MP, Kent SBH, Aebersold R, Barry RA, Tempst P, Teplow DB, Hood LE, Prusiner SB, Weissmann C: A cellular gene encodes scrapie PrP 27–30 protein. Cell 1985;40:735–746.
6   Liao YJ, Lebo RV, Clawson GA, Smuckler EA: Human prion protein cDNA: Molecular cloning, chromosomal mapping, and biological implications. Science 1986;233:364–367.
7   Basler K, Oesch B, Scott M, Westaway D, Wälchli M, Groth DF, McKinley MP, Prusiner SB, Weissmann C: Scrapie and cellular PrP isoforms are encoded by the same chromosomal gene. Cell 1986;46:417–428.
8   Lee IY, Westaway D, Smit AF, Wang K, Seto J, Chen L, Acharya C, Ankener M, Baskin D, Cooper C, Yao H, Prusiner SB, Hood LE: Complete genomic sequence and analysis of the prion protein gene region from three mammalian species. Genome Res 1998;8:1022–1037.
9   Chesebro B, Race R, Wehrly K, Nishio J, Bloom M, Lechner D, Bergstrom S, Robbins K, Mayer L, Keith JM, Garon C, Haase A: Identification of scrapie prion protein-specific mRNA in scrapie-infected and uninfected brain. Nature 1985;315:331–333.
10 Hegde RS, Mastrianni JA, Scott MR, DeFea KA, Tremblay P, Torchia M, DeArmond SJ, Prusiner SB, Lingappa VR: A transmembrane form of the prion protein in neurodegenerative disease. Science 1998;279:827–834.
11 Wopfner F, Weidenhofer G, Schneider R, von Brunn A, Gilch S, Schwarz TF, Werner T, Schatzl HM: Analysis of 27 mammalian and 9 avian PrPs reveals high conservation of flexible regions of the prion rotein. J Mol Biol 1999;289:1163–1178.
12 Bueler H, Fischer M, Lang Y, Bluethmann H, Lipp HP, DeArmond SJ, Prusiner SB, Aguet M, Weissmann C: Normal development and behaviour of mice lacking the neuronal cell-surface PrP protein. Nature 1992;356:577–582.
13 Sakaguchi S, Katamine S, Nishida N, Moriuchi R, Shigematsu K, Sugimoto T, Nakatani A, Kataoka Y, Houtani T, Shirabe S, Okada H, Hasegawa S, Miyamoto T, Noda T: Loss of cerebellar Purkinje cells in aged mice homozygous for a disrupted PrP gene. Nature 1996;380:528–531.
14 Moore RC, Lee IY, Silverman GL, Harrison PM, Strome R, Heinrich C, Karunaratne A, Pasternak SH, Chishti MA, Liang Y, Mastrangelo P, Wang K, Smit AF, Katamine S, Carlson GA, Cohen FE, Prusiner SB, Melton DW, Tremblay P, Hood LE, Westaway D: Ataxia in prion protein (PrP)-deficient mice is associated with upregulation of the novel PrP-like protein doppel. J Mol Biol 1999; 292:797–817.

15   Brown DR, Schmidt B, Kretzschmar HA: Role of microglia and host prion protein in neurotoxicity of a prion protein fragment. Nature 1996;380:345–347.

16   White AR, Collins SJ, Maher F, Jobling MF, Stewart LR, Thyer JM, Beyreuther K, Masters CL, Cappai R: Prion protein-deficient neurons reveal lower glutathione reductase activity and increased susceptibility to hydrogen peroxide toxicity. Am J Pathol 1999;155:1723–1730.

17   Kuwahara C, Takeuchi AM, Nishimura T, Haraguchi K, Kubosaki A, Matsumoto Y, Saeki K, Yokoyama T, Itohara S, Onodera T: Prions prevent neuronal cell-line death. Nature 1999;400: 225–226.

18   Collinge J, Whittington MA, Sidle KCL, Smith CJ, Palmer MS, Clarke AR, Jefferys JGR: Prion protein is necessary for normal synaptic function. Nature 1994;370:295–297.

19   Tagliavini F, Prelli F, Verga L, Giaccone G, Sarma R, Gorevic P, Ghetti B, Passerini F, Ghibaudi E, Forloni G, Salmona M, Bugiani O, Frangione B: Synthetic peptides homologous to prion protein residues 106–147 form amyloid-like fibrils in vitro. Proc Natl Acad Sci USA 1993;90:9678–9682.

20   Forloni G, Angeretti N, Chiesa R, Monzani E, Salmona M, Bugiani O, Tagliavini F: Neurotoxicity of a prion protein fragment. Nature 1993;362:543–546.

21   Brown DR, Herms J, Kretzschmar HA: Mouse cortical cells lacking cellular PrP survive in culture with a neurotoxic PrP fragment. Neuroreport 1994;5:2057–2060.

22   Riek R, Hornemann S, Wider G, Billeter M, Glockshuber R, Wuthrich K: NMR structure of the mouse prion protein domain PrP(121–231). Nature 1996;382:180–182.

23   Donne DG, Viles JH, Groth D, Mehlhorn I, James TL, Cohen FE, Prusiner SB, Wright PE, Dyson HJ: Structure of the recombinant full-length hamster prion protein PrP(29–231): The N terminus is highly flexible. Proc Natl Acad Sci USA 1997;94:13452–13457.

24   James TL, Liu H, Ulyanov NB, Farr-Jones S, Zhang H, Donne DG, Kaneko K, Groth D, Mehlhorn I, Prusiner SB, Cohen FE: Solution structure of a 142-residue recombinant prion protein corresponding to the infectious fragment of the scrapie isoform. Proc Natl Acad Sci USA 1997;94:10086–10091.

25   Riek R, Hornemann S, Wider G, Glockshuber R, Wuthrich K: NMR characterization of the full-length recombinant murine prion protein, mPrP(23–231). FEBS Lett 1997;413:282–288.

26   Zahn R, Liu A, Luhrs T, Riek R, von Schroetter C, Lopez Garcia F, Billeter M, Calzolai L, Wider G, Wuthrich K: NMR solution structure of the human prion protein. Proc Natl Acad Sci USA 2000;97:145–150.

27   Selvaggini C, De Gioia L, Cantu L, Ghibaudi E, Diomede L, Passerini F, Forloni G, Bugiani O, Tagliavini F, Salmona M: Molecular characteristics of a protease-resistant, amyloidogenic and neurotoxic peptide homologous to residues 106–126 of the prion protein. Biochem Biophys Res Commun 1993;194:1380–1386.

28   De Gioia L, Selvaggini C, Ghibaudi E, Diomede L, Bugiani O, Forloni G, Tagliavini F, Salmona M: Conformational polymorphism of the amyloidogenic and neurotoxic peptide homologous to residues 106–126 of the prion protein. J Biol Chem 1994;269:7859–7862.

29   Jobling MF, Stewart LR, White AR, McLean C, Friedhuber A, Maher F, Beyreuther K, Masters CL, Barrow CJ, Collins SJ, Cappai R: The hydrophobic core sequence modulates the neurotoxic and secondary structure properties of the prion peptide 106–126. J Neurochem 1999;73:1557–1565.

30   Hilbich C, Kisters-Woike B, Reed J, Masters CL, Beyreuther K: Substitutions of hydrophobic amino acids reduce the amyloidogenicity of Alzheimer's disease βA4 peptides. J Mol Biol 1992;228:460–473.

31   Pike CJ, Burdick D, Walenciwicz AJ, Glabe CG, Cotman CW: Neurodegeneration induced by β-amyloid peptides in vitro: The role of peptide assembly state. J Neurosci 1993;13:1676–1687.

32   Pike CJ, Walencewicz-Wasserman AJ, Kosmoski J, Cribbs DH, Glabe CG, Cotman CW: Structure-activity analyses of β-amyloid peptides: Contributions of the B25–35 region to aggregation and neurotoxicity. J Neurochem 1995;64:253–265.

33   Palmer MS, Mahal SP, Campbell TA, Hill AF, Sidle KC, Laplanche JL, Collinge J: Deletions in the prion protein gene are not associated with CJD. Hum Molec Genet 1993;2:541–544.

34   Hornshaw MP, McDermott JR, Candy JM, Lakey JH: Copper binding to the N-terminal tandem repeat region of mammalian and avian prion protein: Structural studies using synthetic peptides. Biochem Biophys Res Commun 1995;214:993–999.

35   Hornshaw MP, McDermott JR, Candy JM: Copper binding to the N-terminal tandem repeat regions of mammalian and avian prion protein. Biochem Biophys Res Commun 1995;207:621–629.

36 Stockel J, Safar J, Wallace AC, Cohen FE, Prusiner SB: Prion protein selectively binds copper(II) ions. Biochemistry 1998;37:7185–7193.

37 Miura T, Horii A, Takeuchi H: Metal-dependent α-helix formation promoted by the glycine-rich octapeptide region of prion protein. FEBS Lett 1996;396:248–252.

38 Brown DR, Qin K, Herms JW, Madlung A, Manson J, Strome R, Fraser PE, Kruck T, von Bohlen A, Schulz-Schaeffer W, Giese A, Westaway D, Kretzschmar H: The cellular prion protein binds copper in vivo. Nature 1997;390:684–687.

39 Viles JH, Cohen FE, Prusiner SB, Goodin DB, Wright PE, Dyson HJ: Copper binding to the prion protein: Structural implications of four identical cooperative binding sites. Proc Natl Acad Sci USA 1999;96:2042–2047.

40 Miura T, Hori-i A, Mototani H, Takeuchi H: Raman spectroscopic study on the copper(II) binding mode of prion octapeptide and its pH dependence. Biochemistry 1999;38:11560–11569.

41 Brown DR, Hafiz F, Glasssmith LL, Wong BS, Jones IM, Clive C, Haswell SJ: Consequences of manganese replacement of copper for prion protein function and proteinase resistance. EMBO J 2000;19:1180–1186.

42 Waggoner DJ, Drisaldi B, Bartnikas TB, Casareno RL, Prohaska JR, Gitlin JD, Harris DA: Brain copper content and cuproenzyme activity do not vary with prion protein expression level. J Biol Chem 2000;275:7455–7458.

43 Wadsworth JD, Hill AF, Joiner S, Jackson GS, Clarke AR, Collinge J: Strain-specific prion-protein conformation determined by metal ions. Nat Cell Biol 1999;1:55–59.

44 Bush A: Metal and neuroscience. Curr Opin Chem Biol 2000;4:184–191.

45 Safar J, Roller PP, Gajdusek DC, Gibbs CJ Jr: Conformational transitions, dissociation, and un-folding of scrapie amyloid (prion) protein. J Biol Chem 1993;268:20276–20284.

46 Safar J, Roller PP, Gajdusek DC, Gibbs CJ Jr: Scrapie amyloid (prion) protein has the conformational characteristics of an aggregated molten globule folding intermediate. Biochemistry 1994;33:8375–8383.

47 Mehlhorn I, Groth D, Stockel J, Moffat B, Reilly D, Yansura D, Willett WS, Baldwin M, Fletterick R, Cohen FE, Vandlen R, Henner D, Prusiner SB: High-level expression and characterization of a purified 142-residue polypeptide of the prion protein. Biochemistry 1996;35:5528–5537.

48 Hornemann S, Korth C, Oesch B, Riek R, Wider G, Wuthrich K, Glockshuber R: Recombinant full-length murine prion protein, mPrP(23–231): Purification and spectroscopic characterization. FEBS Lett 1997;413:277–281.

49 Huang Z, Gabriel J-M, Baldwin MA, Fletterick RJ, Prusiner SB, Cohen FE: Proposed three-dimensional structure for the cellular prion protein. Proc Natl Acad Sci USA 1994;91:7139–7143.

50 Liu H, Farr-Jones S, Ulyanov NB, Llinas M, Marqusee S, Groth D, Cohen FE, Prusiner SB, James TL: Solution structure of Syrian hamster prion protein rPrP(90–231). Biochemistry 1999;38:5362–5377.

51 Billeter M, Riek R, Wider G, Hornemann S, Glockshuber R, Wuthrich K: Prion protein NMR structure and species barrier for prion diseases. Proc Natl Acad Sci USA 1997;94:7281–7285.

52 Hsiao KK, Groth D, Scott M, Yang S-L, Serban H, Rapp D, Foster D, Torchia M, DeArmond SJ, Prusiner SB: Serial transmission in rodents of neurodegeneration from transgenic mice expressing mutant prion protein. Proc Natl Acad Sci USA 1994;91:9126–9130.

53 Telling GC, Haga T, Torchia M, Tremblay P, DeArmond SJ, Prusiner SB: Interactions between wild-type and mutant prion proteins modulate neurodegeneration transgenic mice. Genes Dev 1996;10:1736–1750.

54 Chiesa R, Drisaldi B, Quaglio E, Migheli A, Piccardo P, Ghetti B, Harris DA: Accumulation of protease-resistant prion protein (PrP) and apoptosis of cerebellar granule cells in transgenic mice expressing a PrP insertional mutation. Proc Natl Acad Sci USA 2000;97:5574–5579.

55 Lehmann S and Harris DA: Mutant and infectious prion proteins display common biochemical properties in cultured cells. J Biol Chem 1996;271:1633–1637.

56 Petersen RB, Parchi P, Richardson SL, Urig CB, Gambetti P: Effect of the D178N mutation and the codon 129 polymorphism on the metabolism of the prion protein. J Biol Chem 1996;271:12661–12668.

57   Swietnicki W, Petersen RB, Gambetti P, Surewicz WK: Familial mutations and the thermodynamic stability of the recombinant human prion protein. J Biol Chem 1998;273:31048–31052.

58   Cappai R, Stewart L, Jobling MF, Thyer JM, White AR, Beyreuther K, Collins SJ, Masters CL, Barrow CJ: Familial prion disease mutation alters the secondary structure of recombinant mouse prion protein: Implications for the mechanism of prion formation. Biochemistry 1999;38:3280–3284.

59   Zhang H, Stockel J, Mehlhorn I, Groth D, Baldwin MA, Prusiner SB, James TL, Cohen FE: Physical studies of conformational plasticity in a recombinant prion protein. Biochemistry 1997; 36:3543–3553.

60   Kocisko DA, Lansbury PT, Caughey B: Partial unfolding and refolding of scrapie-associated prion protein – Evidence for a critical 16-kDa C-terminal domain. Biochemistry 1996;35:13434–13442.

61   Peretz D, Williamson RA, Matsunaga Y, Serban H, Pinilla C, Bastidas RB, Rozenshteyn R, James TL, Houghten RA, Cohen FE, Prusiner SB, Burton DR: A conformational transition at the N terminus of the prion protein features in formation of the scrapie isoform. J Mol Biol 1997;273: 614–622.

62   Muramoto T, Scott M, Cohen FE, Prusiner SB: Recombinant scrapie-like prion protein of 106 amino acids is soluble. Proc Natl Acad Sci USA 1996;93:15457–15462.

63   Safar J, Wille H, Itri V, Groth D, Serban H, Torchia M, Cohen FE, Prusiner SB: Eight prion strains have PrP(Sc) molecules with different conformations. Nat Med 1998;4:1157–1165.

64   Hosszu LL, Baxter NJ, Jackson GS, Power A, Clarke AR, Waltho JP, Craven CJ, Collinge J: Structural mobility of the human prion protein probed by backbone hydrogen exchange. Nat Struct Biol 1999;6:740–743.

65   Swietnicki W, Petersen R, Gambetti P, Surewicz WK: pH-dependent stability and conformation of the recombinant human prion protein PrP(90–231). J Biol Chem 1997;272:27517–27520.

66   Morillas M, Swietnicki W, Gambetti P, Surewicz WK: Membrane environment alters the conformational structure of the recombinant human prion protein. J Biol Chem 1999;274:36859–36865.

67   Wildegger G, Liemann S, Glockshuber R: Extremely rapid folding of the C-terminal domain of the prion protein without kinetic intermediates. Nat Struct Biol 1999;6:550–553.

68   Jackson GS, Hosszu LL, Power A, Hill AF, Kenney J, Saibil H, Craven CJ, Waltho JP, Clarke AR, Collinge J: Reversible conversion of monomeric human prion protein between native and fibrilogenic conformations. Science 1999;283:1935–1937.

69   Nguyen JT, Inouye H, Baldwin MA, Fletterick RJ, Cohen FE, Prusiner SB, Kirschner DA: X-ray diffraction of scrapie prion rods and PrP peptides. J Mol Biol 1995;252:412–422.

70   Wille H, Prusiner SB: Ultrastructural studies on scrapie prion protein crystals obtained from reverse micellar solutions. Biophys J 1999;76:1048–1062.

Dr. Roberto Cappai, Department of Pathology, The University of Melbourne,
Parkville, Victoria 3010 (Australia)
Tel. +61 3 8344 5868, Fax +61 3 8344 4004, E-Mail r.cappai@pathology.unimelb.edu.au

Rabenau HF, Cinatl J, Doerr HW (eds): Prions. A Challenge for Science,
Medicine and Public Health System. Contrib Microb. Basel, Karger, 2001, vol 7, pp 48–57

....

Strain Variations and Species Barriers

Andrew F. Hill, John Collinge

MRC Prion Unit, Department of Neurogenetics, Imperial College School of
Medicine at St. Mary's, London, UK

Prion diseases or transmissible spongiform encephalopathies are a group of neurodegenerative disorders affecting both humans and animals. These diseases include scrapie in sheep, bovine spongiform encephalopathy (BSE) in cattle, Creutzfeldt-Jakob disease (CJD), Gerstmann-Sträussler-Sheinker disease, fatal familial insomnia and kuru. Prion diseases are transmissible by inoculation, have long incubation periods and share common histological features. One of the central features of prion disease is the conversion of the normal cellular form of the host-encoded prion protein (PrP^C) to an abnormal isoform designated PrP^{Sc}. This conversion occurs post-translationally and is thought to involve a conformational change rather then a covalent modification. PrP^{Sc} may be distinguished from PrP^C by its insolubility in detergent and partial resistance to protease degradation. Prion diseases in humans may be inherited through germ-line mutations in the human prion gene (*PRNP*), acquired through inoculation (including dietary exposure) or caused by rare events which convert PrP^C molecules to PrP^{Sc}. Much data exists to suggest that the principal component of the transmissible agent, or prion, is an abnormal isoform of PrP^C and forms the basis of the protein-only hypothesis of prion propagation. This hypothesis suggests that PrP^{Sc} replicates itself by recruiting PrP^C molecules and inducing a conformational change, resulting in the accumulation of further PrP^{Sc} which may in turn convert more of the cellular isoform [1]. Ablation of the prion gene in transgenic mice (termed *Prnp*$^{0/0}$ mice) supports this hypothesis, with the animals being both resistant to experimental scrapie and failing to propagate prion infectivity, firmly supporting the role of PrP in these diseases [2–4].

Species Barriers

Prion diseases are characterized by their transmissibility to experimental animals. This has been shown by the transmission of kuru to chimpanzees [5], and CJD [6] or GSS [7] to primates. Experimentally, prion diseases are transmitted by intracerebral inoculation, but they may also be transmitted via intraperitoneal and oral routes.

Traditionally, transmission of prion diseases between species is difficult, requiring the exposure of large numbers of animals to obtain illness in only a few individuals. In addition, incubation periods are lengthened in the new host species and often may approach the natural life-span of the inoculated animal. These incubation periods shorten on second passage in the new host species and reduce to a stable level on subsequent passages. This phenomenon is known as the 'species barrier' [8] and may relate to many factors, such as the strain of prion inoculated, the difference in primary sequence of the prion protein between donor and host species, and the route of inoculation.

One species barrier which has been extensively examined is that between hamsters and mice. Kimberlin and Marsh [9] described a hamster model of scrapie in which infected hamsters became sick 60–70 days after inoculation, the shortest reported incubation period for rodent models of scrapie. The hamster prions were found to be pathogenic for hamsters and not mice, suggesting a substantial species barrier between the two species. Advances in transgenic animal technology have proved invaluable in exploring this phenomenon, and abrogation of the species barrier was demonstrated in transgenic mice overexpressing PrP sequences from the donor species. It was shown that transgenic mice overexpressing Syrian hamster PrP transgenes showed no species barrier when infected with hamster prions, succumbing to illness at around 75 days after inoculation [10]. The wild-type littermates of the transgenic mice remained symptom free at over 500 days after inoculation. Additionally, the transgenic mice were found to harbour high levels of hamster and not mouse PrP^{Sc}, which was found on passage to be pathogenic for hamsters and not mice [11]. Further experiments established that transgenic mice generated on a $Prnp^{0/0}$ background reduced the species barrier effect further, with a reduction of incubation period concomitant with removal of endogenous mouse PrP [12]. This led to the notion that primary PrP sequence is one of the main factors involved in the species barrier. Experiments using transgenic mice overexpressing human PrP genes have also been used to abrogate the species barrier between mice and humans [13–16]. Transgenic mice expressing human PrP have consistent incubation periods of between 180 and 220 days when inoculated with human prions from classical CJD cases. In these mice, it has been possible to transmit a variety of human prion diseases including inherited, acquired, sporadic, variant CJD (vCJD), and FFI [14, 16–18]. These mice have also been useful in ascertaining the risk of humans to prion diseases of other animals, in particular BSE from cattle. It was found that at prolonged incubation periods (>500 days) some of the BSE infected 'humanized' mice (which encode valine at codon 129, a common polymorphism in humans) showed clinical symptoms and neuropathology consistent with prion disease [14]. That these mice succumbed at greatly increased incubation periods with less than a 100% attack rate suggests that a substantial species barrier is present for this human PrP genotype. However, cases of vCJD have only been observed in humans who are methionine homozygous at codon 129, so it will be important to assess the susceptibility of transgenic mice expressing this genotype to infection with BSE.

Another study on the transmissibility of human prion diseases to transgenic mice suggested that expression of human PrP genes on a $Prnp^{0/0}$ background was necessary for transmission of human prions and that the presence of endogenous mouse PrP interfered with ablation of the species barrier [13]. Similar results were seen in experiments with transgenic mice expressing bovine PrP genes where the mice bred on a $Prnp^{0/0}$ background had much shorter incubation periods than those containing endogenous mouse PrP genes [19]. As the primary sequence alone was not the sole determinant of the species barrier, an as yet unknown factor termed 'protein X' was hypothesized to have an effect on the species barrier phenomenon.

In exploring the role of the primary PrP sequence on the species barrier, several transgenic mice were generated containing chimeric mouse/hamster PrP genes. One such construct, where the central third of the mouse open reading frame is replaced with the corresponding section of hamster sequence (called MH2M), when expressed in transgenic mice generates prions with an artificial host range [20]. Such transgenic mice are susceptible to infection with both mouse and hamster prions, and the PrP^{Sc} formed in these animals is chimeric in origin. These chimeric prions have been shown to be transmissible to both mice and hamsters and have proved useful in examining some of the many strains of mouse scrapie in the hamster by passage through this intermediate host, which shows no appreciable species barrier to either source of prions [21]. Similar experiments have been performed in transgenic mice expressing mouse-human [15] and mouse-bovine [19] chimeric PrP transgenes (MHu2M and MBo2M, respectively). Interestingly, the MBo2M transgenic mice were not sensitive to inoculation with bovine prions and it was presumed that differences in amino acid residues 180–205 between mouse and bovine PrP genes may be responsible, however this issue remains unresolved.

Importantly, there are other examples where the primary PrP sequence alone cannot explain such species barriers, perhaps best illustrated by the natural transmission of BSE to a wide variety of hosts, all with different PrP primary structures. It has been clear for many years that prion strain type (see below) has a crucial effect on species barriers. A striking example of the strain effect to species barriers has been provided by analysis of BSE prions. While classical CJD prions, propagated in humans expressing wild-type human PrP, transmit highly efficiently to mice expressing only human PrP with transmission characteristics consistent with complete absence of a species barrier [14], vCJD prions, also propagated in humans expressing wild-type PrP of identical primary structure, have transmission properties completely distinct from other human prions (as assessed either in transgenic or wild-type mice) but indistinguishable from those of cattle BSE [17, 22] and consistent with the presence of a transmission barrier.

In addition, recent studies have suggested that mice inoculated with hamster prions may replicate or harbour the infectious agent without showing any clinical signs, and may transmit the disease when repassaged back into hamsters [23]. The incubation periods of disease in the hamsters inoculated with the mouse-passaged material were longer than that of the original inoculum, which may indicate persistence of the agent in the brains of the mice. However, it is possible that prion replication has occurred in the mice generating a strain of prions which has different transmission properties when inoculated into hamsters. It now seems preferable to use the term 'transmission barrier' to indicate the difficulty in transmitting prions from one host to another [24].

Strain Characteristics

While there are several lines of evidence to support the protein-only hypothesis of prion propagation, the existence of distinct 'strains' or isolates which can be stably passaged in inbred mice of the same genotype has been a challenge to accommodate in this model. Prion strains can be distinguished by their different incubation periods and patterns of neuropathology when passaged in inbred mice. Three hypotheses have been proposed to explain the existence of prion strains. The first is the virus or virino hypothesis which assumes a mutation in an agent-specific DNA or RNA genome. However, no direct evidence for such an agent-specific genome has been produced. The second hypothesis tries to bridge the gap between the virus and protein-only hypotheses. Weissmann's [25] unified hypothesis states that while the protein is sufficient for infectivity, some small nucleic acid confers strain properties on the protein. The third is the protein-only hypothesis where the protein itself encodes the information required to determine the strain phenotype.

Evidence that strain specificity is encoded by PrP itself, and not a nucleic acid, was provided by studies on two distinct strains of transmissible mink encephalopathy (TME). Termed hyper (HY) and drowsy (DY), these strains can be serially passaged in hamsters. The strains can be distinguished by physicochemical properties of the PrP^{Sc} deposited in the brains of the infected hamsters. After treatment with proteinase K, strain-specific migration patterns can be seen on Western blots, with DY PrP^{Sc} being more protease sensitive than HY PrP^{Sc} and producing different banding patterns [26]. The banding patterns are caused by different N-terminal cleavage sites for proteinase K which suggest the strains represent different PrP^{Sc} conformations, which has recently been supported by infra-red spectroscopic studies [27]. Maintenance of these TME strains has also been shown in an in vitro conversion model when hamster PrP^{C} is mixed with HY or DY PrP^{Sc}, which also supports the view that prion strains involve different PrP conformers [28, 29].

Several human PrP^{Sc} types have recently been identified which are associated with different clinicopathological phenotypes of CJD [16, 30, 31]. These types are distinguished by different fragment sizes seen after limited proteinase K digestion suggesting different conformations of PrP^{Sc}. These types can be further classified by the ratio of the three PrP bands seen after protease digestion, representing di-, mono- and un-glycosylated fragments of PrP^{Sc}. PrP^{Sc} conformation and glycosylation are therefore plausible candidates as forming the molecular basis of prion strain diversity. However, it is crucial to then determine whether such biochemical properties fulfil the biological characteristics of strains, that is that they are maintained upon transmission to experimental animals of both the same and different species. This was confirmed with studies using transgenic mice overexpressing human PrP (which show no species barrier to infection with human prions [14]) which demonstrated that both the fragment sizes and glycoform ratios can be maintained upon transmission [16, 17]. Additionally, transmission of human CJD prions and BSE to wild-type mice also results in maintenance of the fragment size and glycoform ratio of the inocula. Classical CJD is associated with PrP^{Sc} types 1–3, while type 4 human PrP^{Sc} is uniquely associated with vCJD and characterized by glycoform ratios which are distinct from those observed in classical CJD [16]. The glycoform ratios in vCJD are similar to those seen in BSE and cases of natural or experimental BSE transmission to a variety of species. The PrP^{Sc} types formed in wild-type mice inoculated with either BSE or vCJD are virtually identical [17]. Coupled with other parameters of transmission, such as the incubation period, numbers of animals succumbing to illness, and traditional lesion-profiling methods provides compelling evidence that these two diseases are caused by the same prion strain [17, 22]. This can be observed readily by plotting the ratios of the glycoforms in a scattergram

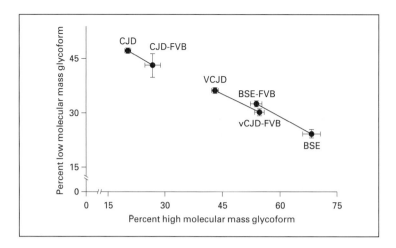

Fig. 1. Scattergram illustrating the glycoform profiles for sporadic CJD, variant CJD and BSE transmitted to wild-type FVB mice. Shown are mean values with SEM plotted as error bars.

(fig. 1). Here it is possible to observe that although the glycoform ratios of vCJD and BSE appear statistically dissimilar, transmission of each of these to a common host shows them to be indistinguishable. Presumably, there are strain-, species- and tissue-specific effects on PrP glycosylation such that, as with classical biological strain typing, comparison of isolates on the same background may be necessary to reveal the strain characteristics most clearly. This is also shown in the glycoform ratios of classical CJD in humans and the transmission of these types to the same lines of wild-type mice. Thus, there are also host-specific factors involved in the molecular basis of prion strains and these could include both the primary PrP structure and host-specific glycosylation. These data strongly support the 'protein only' hypothesis of infectivity and suggest that strain variation is encoded by a combination of PrP conformation and glycosylation. Further evidence to suggest that strain-specific information is encoded in the conformation of PrP^{Sc} comes from transmission of two inherited forms of human prion disease to transgenic mice expressing a chimeric mouse/human prion gene. Here it was shown that the conformation of the inocula (as measured by fragment size after protease cleavage) was maintained upon transmission to the transgenic animal. Furthermore, the incubation periods and patterns of neuropathology were found to differ between the two inoculum and illustrates that the chimeric PrP which is converted in these transgenic mice may adopt two distinct conformations which lead to different disease profiles [32].

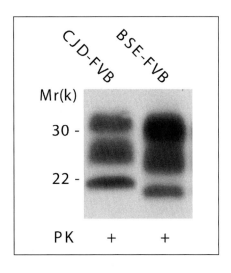

Fig. 2. Western blot illustrating the different proteinase K fragment sizes and glycoform profiles observed in wild-type FVB mice inoculated with either sporadic CJD (lane 1) or BSE (lane 2).

As PrP glycosylation occurs before conversion to PrP^{Sc}, the different glycoform ratios may represent selection of particular PrP^C glycoforms by PrP^{Sc} of different conformations with inoculated prions preferentially recruiting and converting particular glycoforms of PrP^C. This effect can be illustrated in the transmission of classical CJD and BSE to wild-type mice where two different glycoform patterns and fragment sizes can be generated from the same pool of PrP^C (fig. 2). According to such a hypothesis, PrP conformation would be the primary determinant of strain type with glycosylation being involved as a secondary process. However, since it is known that different cell types may glycosylate proteins differently, PrP^{Sc} glycosylation patterns may provide a substrate for the neuropathological targeting that distinguishes different prion strains. Particular PrP^{Sc} glycoforms may replicate most favourably in neuronal populations with a similar PrP glycoform expressed on the cell surface. Such targeting could also explain the different incubation periods which also discriminate strains, targeting of more critical brain regions, or regions with higher levels of PrP expression, producing shorter incubation periods. Further supportive evidence for the involvement of PrP glycosylation in prion strain propagation has come from the study of transgenic mice expressing PrP with mutations interfering with N-linked glycosylation [33]. These mutations led to aberrant distribution of PrP^C which also affected the ability of these mice to be infected with prions. Mutation of the first glycosylation consensus sequence resulted in a restricted pattern of PrP^C expression which did not support PrP^{Sc} replication. Interestingly, mutation of the second glycosylation site resulted in a wider expression pattern of PrP^C, which supported prion replication; however

---

the incubation period for infection with hamster prions in these mice was over 500 days. Differences in the relative spatial expression of the glycoform mutants however compromise the use of this model in directly evaluating the precise role of glycosylation as a determinant of strain variation. Indeed, it has been established that there may be considerable heterogeneity among the carbohydrate trees attached to PrP^C (selected from over 400 distinct structures) [34] and as a consequence it will be an extremely complex task to determine whether particular carbohydrates are preferentially involved in the replication of distinct prion strains. Evidence that this may be the case is exemplified by the apparent superimposition of strain- and tissue-specific effects on PrP glycosylation seen in vCJD tonsil PrP^{Sc}, which differs in the proportion of the PrP glycoforms from that seen in vCJD brain [35]. Transmission of PrP^{Sc} from peripheral tissues will be useful in trying to assess the contribution of glycosylation in conferring strain-specific properties.

While the function of PrP still remains unknown, its role in the development of prion disease is firmly established through transgenic mouse experiments. With $Prnp^{0/0}$ mice being resistant to infection with prions, a series of transgenic mice containing N-terminal deletions of PrP have been generated and assessed for the ability to propagate prion disease and probe structure/ function studies of this protein. Transgenic mice expressing PrP with deletions up to residue 93 (on a $Prnp^{0/0}$ background) are phenotypically normal, develop scrapie when inoculated with mouse prions, and generate protease-resistant truncated PrP, which may be propagated [36]. Interestingly, these animals lack the octapeptide repeat region, and as insertions of additional repeat units are pathogenic in humans [37], raise some interesting questions as to the functional significance of this region of the protein, which has been shown by solution NMR spectroscopy to have no discernible tertiary structure [38, 39]. Another transgenic mouse which contained deletion of the first α-helix , β-strand and part of helix 2 (PrPΔ23-88Δ141-176) were also susceptible to infection of prions and were able to propagate the infectivity generated [40]. Another set of experiments using transgenic mice over-expressing N-terminal deletions extending beyond mouse residues 106–121 and 134 develop an ataxic syndrome between 3 and 8 weeks of age. That this phenotype could be rescued by introducing a single wild-type PrP allele suggests that this phenomenon is not simply due to the expression of the truncated PrP protein, but rather its interference in an as yet unknown signalling pathway [36].

The use of transgenic mice in research into prion diseases have been an invaluable resource in developing our understanding of the molecular basis of species barriers and prion strain variation. While the exact mechanism of conversion from PrP^C to PrP^{Sc} is not yet understood, these mouse models provide a means for examining the biological role of PrP and may prove to be useful in determining the function of this protein and its role in these neurodegenerative diseases.

References

1 Prusiner SB: Novel proteinaceous infectious particles cause scrapie. Science 1982;216:136–144.
2 Bueler H, Fischer M, Lang Y, Bluethmann H, Lipp H-P, DeArmond SJ, Prusiner SB, Aguet M, Weissmann C: Normal development and behaviour of mice lacking the neuronal cell-surface PrP protein. Nature 1992;356:577–582.
3 Bueler H, Aguzzi A, Sailer A, Greiner RA, Autenried P, Aguet M, Weissmann C: Mice devoid of PrP are resistant to scrapie. Cell 1993;73:1339–1347.
4 Sailer A, Bueler H, Fischer M, Aguzzi A, Weissmann C: No propagation of prions in mice devoid of PrP. Cell 1994;77:967–968.
5 Gajdusek DC, Gibbs CJ, Alpers M: Experimental transmission of a Kuru-like syndrome to chimpanzees. Nature 1966;209:794–796.
6 Gajdusek DC, Gibbs CJ: Transmission of two subacute spongiform encephalopathies of man (kuru and Creutzfeldt-Jakob disease) to New World monkeys. Nature 1971;230:588–591.
7 Gibbs CJ, Gajdusek DC: Experimental subacute spongiform virus encephalopathies in primates and other laboratory animals. Science 1973;182:67–68.
8 Pattison IH: Experiments with scrapie with special reference to the nature of the agent and the pathology of the disease; in Gajdusek CJ, Gibbs CJ, Alpers MP (eds): Slow, Latent and Temperate Virus Infections. Washington, US Government Printing, 1965, NINDB Monograph 2, pp 249–257.
9 Marsh RF, Kimberlin RH: Comparison of scrapie and transmissible mink encephalopathy in hamsters. II. Clinical signs, pathology, and pathogenesis. J Infect Dis 1975;131:104–110.
10 Scott M, Foster D, Mirenda C, Serban D, Coufal F, Wälchli M, Torchia M, Groth D, Carlson G, DeArmond SJ, Westaway D, Prusiner SB: Transgenic mice expressing hamster prion protein produce species- specific scrapie infectivity and amyloid plaques. Cell 1989;59:847–857.
11 Prusiner SB, Scott M, Foster D, Pan KM, Groth D, Mirenda C, Torchia M, Yang SL, Serban D, Carlson GA, Hoppe PC, Westaway D, DeArmond SJ: Transgenetic studies implicate interactions between homologous PrP isoforms in scrapie prion replication. Cell 1990;63:673–686.
12 Bueler H, Raeber A, Sailer A, Fischer M, Aguzzi A, Weissmann C: High prion and PrPSc levels but delayed onset of disease in scrapie-inoculated mice heterozygous for a disrupted PrP gene. Mol Med 1994;1:19–30.
13 Telling GC, Scott M, Mastrianni J, Gabizon R, Torchia M, Cohen FE, DeArmond SJ, Prusiner SB: Prion propagation in mice expressing human and chimeric PrP transgenes implicates the interaction of cellular PrP with another protein. Cell 1995;83:79–90.
14 Collinge J, Palmer MS, Sidle KCL, Hill AF, Gowland I, Meads J, Asante E, Bradley R, Doey LJ, Lantos PL: Unaltered susceptibility to BSE in transgenic mice expressing human prion protein. Nature 1995;378:779–783.
15 Telling GC, Scott M, Hsiao KK, Foster D, Yang S-L, Torchia M, Sidle KCL, Collinge J, DeArmond SJ, Prusiner SB: Transmission of Creutzfeldt-Jakob disease from humans to transgenic mice expressing chimeric human-mouse prion protein. Proc Natl Acad Sci USA 1994;91:9936–9940.
16 Collinge J, Sidle KCL, Meads J, Ironside J, Hill AF: Molecular analysis of prion strain variation and the aetiology of 'new variant' CJD. Nature 1996;383:685–690.
17 Hill AF, Desbruslais M, Joiner S, Sidle KCL, Gowland I, Doey L, Lantos P, Collinge J: The same prion strain causes vCJD and BSE. Nature 1997;389:448–450.
18 Collinge J, Palmer MS, Sidle KCL, Gowland I, Medori R, Ironside J, Lantos PL: Transmission of fatal familial insomnia to laboratory animals. Lancet 1995;346:569–570.
19 Scott MR, Safar J, Telling G, Nguyen O, Groth D, Torchia M, Koehler R, Tremblay P, Walther D, Cohen FE, DeArmond SJ, Prusiner SB: Identification of a prion protein epitope modulating transmission of bovine spongiform encephalopathy prions to transgenic mice. Proc Natl Acad Sci USA 1997; 94:14279–14284.

20 Scott M, Groth D, Foster D, Torchia M, Yang SL, DeArmond SJ, Prusiner SB: Propagation of prions with artificial properties in transgenic mice expressing chimeric PrP genes. Cell 1993;73:979–988.

21 Scott MR, Groth D, Tatzelt J, Torchia M, Tremblay P, DeArmond SJ, Prusiner SB: Propagation of prion strains through specific conformers of the prion protein. J Virol 1997;71:9032–9044.

22 Bruce ME, Will RG, Ironside JW, McConnell I, Drummond D, Suttie A, McCardle L, Chree A, Hope J, Birkett C, Cousens S, Fraser H, Bostock CJ: Transmissions to mice indicate that 'new variant' CJD is caused by the BSE agent. Nature 1997;389:498–501.

23 Race R, Chesebro B: Scrapie infectivity found in resistant species. Nature 1998;392:770.

24 Collinge J: Variant Creutzfeldt-Jakob disease. Lancet 1999;354:317–323.

25 Weissmann C: A 'unified theory' of prion propagation. Nature 1991;352:679–683.

26 Marsh RF, Bessen RA, Lehmann S, Hartsough GR: Epidemiological and experimental studies on a new incident of transmissible mink encephalopathy. J Gen Virol 1991;72:589–594.

27 Caughey B, Raymond GJ, Bessen RA: Strain-dependant differences in β-sheet conformations of abnormal prion protein. J Biol Chem 1998;273:32230–32225.

28 Bessen RA, Kocisko DA, Raymond GJ, Nandan S, Lansbury PT, Caughey B: Non-genetic propagation of strain-specific properties of scrapie prion protein. Nature 1995;375:698–700.

29 Bessen RA, Raymond GJ, Caughey B: In situ formation of protease-resistant prion protein in transmissible spongiform encephalopathy-infected brain slices. J Biol Chem 1997;272:15227–15231.

30 Parchi P, Castellani R, Capellari S, Ghetti B, Young K, Chen SG, Farlow M, Dickson DW, Sims AAF, Trojanowski JQ, Petersen RB, Gambetti P: Molecular basis of phenotypic variability in sporadic Creutzfeldt-Jakob disease. Ann Neurol 1996;39:767–778.

31 Wadsworth JDF, Hill AF, Joiner S, Jackson GS, Clarke AR, Collinge J: Strain-specific prion-protein conformation determined by metal ions. Nat Cell Biol 1999;1:55–59.

32 Telling GC, Parchi P, DeArmond SJ, Cortelli P, Montagna P, Gabizon R, Mastrianni J, Lugaresi E, Gambetti P, Prusiner SB: Evidence for the conformation of the pathologic isoform of the prion protein enciphering and propagating prion diversity. Science 1996;274:2079–2082.

33 DeArmond SJ, Sánchez H, Yehiely F, Qiu Y, Ninchak-Casey A, Daggett V, Camerino AP, Cayetano J, Rogers M, Groth D, Torchia M, Tremblay P, Scott MR, Cohen FE, Prusiner SB: Selective neuronal targeting in prion disease. Neuron 1997;19:1337–1348.

34 Endo T, Groth D, Prusiner SB, Kobata A: Diversity of oligosaccharide structures linked to asparagines of the scrapie prion protein. Biochemistry 1989;28:8380–8388.

35 Hill AF, Butterworth RJ, Joiner S, Jackson G, Rossor MN, Thomas DJ, Frosh A, Tolley N, Bell JE, Spencer M, King A, Al-Sarraj S, Ironside JW, Lantos PL, Collinge J: Investigation of variant Creutzfeldt-Jakob disease and other human prion diseases with tonsil biopsy samples. Lancet 1999;353:183–189.

36 Shmerling D, Hegyi I, Fischer M, Blattler T, Brandner S, Gotz J, Rulicke T, Flechsig E, Cozzio A, von-Mering C, Hangartner C, Aguzzi A, Weissmann C: Expression of amino-terminally truncated PrP in the mouse leading to ataxia and specific cerebellar lesions. Cell 1998;93:203–214.

37 Collinge J: Human prion diseases and bovine spongiform encephalopathy (BSE). Hum Mol Genet 1997;6:699–705.

38 Riek R, Hornemann S, Wider G, Glockshuber R, Wüthrich K: NMR characterisation of the full-length recombinant murine prion protein mPrP(23–231). Febs Lett 1997;413:282–288.

39 Donne DG, Viles JH, Groth D, Mehlhorn I, James TL, Cohen FE, Prusiner SB, Wright PE, Dyson HJ: Structure of the recombinant full-length hamster prion protein PrP(29–231): The N terminus is highly flexible. Proc Natl Acad Sci USA 1997;94:13452–13457.

40 Supattapone S, Bosque P, Muramoto T, Wille H, Aagaard C, Peretz D, Nguyen HOB, Heinrich C, Torchia M, Safar J, Cohen FE, DeArmond SJ, Prusiner SB, Scott M: Prion protein of 106 residues creates an artificial transmission barrier for prion replication in transgenic mice. Cell 1999;96:869–878.

Andrew F. Hill, Department of Pathology, The University of Melbourne,
Parkville, Victoria 3052 (Australia)
Tel. +61 3 8344 3995, Fax +61 3 8344 4004, E-Mail a.hill@unimelb.edu.au

Rabenau HF, Cinatl J, Doerr HW (eds): Prions. A Challenge for Science,
Medicine and Public Health System. Contrib Microb. Basel, Karger, 2001, vol 7, pp 58–67

....

Resistance of Transmissible Spongiform Encephalopathy Agents to Decontamination

D. M. Taylor

Neuropathogenesis Unit, Institute for Animal Health, Edinburgh, UK

The transmissible spongiform encephalopathies (TSEs) include scrapie in sheep, bovine spongiform encephalopathy (BSE), and Creutzfeldt-Jakob disease (CJD) of humans. In TSEs, a normal host-protein (PrP^C) is converted to a disease-specific form (PrP^{Sc}) as a consequence of infection. PrP^{Sc} resists proteolytic digestion and forms pathological deposits, particularly within the central nervous system (CNS) where this is usually accompanied by spongiform encephalopathy. One theory is that TSE agents are simply PrP^{Sc} (perhaps accompanied by other host-specific proteins) which acts as a template for the conversion of PrP^C to PrP^{Sc} [1, 2]. Although there is no general disagreement with the idea that PrP^{Sc} is at least a component of such agents, there is an opinion [3–6] that the protein-only ('prion') hypothesis cannot explain (a) the variety of phenotypic characteristics of different strains of scrapie agent in mice of the same *PrP* genotype, or (b) the phenotypic stability of the BSE agent in mice, regardless of whether transmission is directly from cattle to mice, or via intermediate species such as kudu, nyala, domestic cats, pigs, sheep, goats, and even humans [7, 8]. Although the incidence of BSE in the UK is declining, concern has been heightened by its putative link with a new variant form of CJD (vCJD) first reported in 1996 [9]. The phenotypic characteristics of this agent in mice are the same as the BSE agent which is unlike any other TSE agent that has yet been characterized [8]. Although TSE agents have not been fully characterized, they are known to have a high degree of resistance to inactivation, which has resulted in accidental transmission. CJD was transmitted accidentally by using inadequately decontaminated neurosurgical equipment [10, 11]. Scrapie was transmitted accidentally to sheep and goats through the survival of this agent

Table 1. Degree of inactivation of TSE agents achieved by various procedures[1]

No detectable infectivity	Significant titre reduction	Little titre reduction
Sodium hypochlorite (16,500 ppm available chlorine)	1 *M* or 2 *M* sodium hydroxide	aldehydes organic solvents hydrogen peroxide
Autoclaving at 121 °C after 1 *M* sodium hydroxide treatment	sodium dichoroisocyanurate (16,500 ppm available chlorine)	phenolic disinfectants chlorine dioxide iodine and iodates
Autoclaving at 121 °C in 1 *M* sodium hydroxide	chaotropes (e.g. guanidine thiocyanate)	peracetic acid protcolytic enzymes microwave irradiation
Boiling in 1 *M* sodium hydroxide	95% formic acid hot 1 *M* hydrochloric acid	UV irradiation gamma irradiation
	autoclaving for 18 min at 134–138 °C	autoclaving after aldehyde, alcohol or dry heat treatments
	autoclaving for 1 h at 132 °C	
	autoclaving at 121 °C in 5% sodium dodecyl sulphate	
	dry heat at > 200 °C	

[1] Based upon data presented in this publication, and in Taylor [22].

in formol-treated vaccines [12, 13]. BSE was transmitted accidentally through the survival of infectivity after the cooking procedures used to manufacture meat and bone meal for inclusion in animal feed [14]; the failure of most of these procedures to completely inactivate BSE and scrapie agents has been demonstrated [15, 16]. It was considered by the late 1980s that a few reliable decontamination procedures for TSE agents had been established. In the UK, the recommended methods were porous-load (PL) autoclaving at 134–138 °C for 18 min [17], or exposure to sodium hypochlorite solution containing 20,000 ppm available chlorine for 1 h [18]. In the USA, gravity-displacement (GD) autoclaving at 132 °C for 1 h, or exposure to 1 *M* sodium hydroxide for 1 h, was preferred [19]. These recommended procedures were adopted worldwide, and have been incorporated into formal recommendations on how to deal with TSE infectivity [e.g. 19, 20]. However, further decontamination studies on BSE and scrapie agents have cast doubt on the reliability of three of these recommended methods. These, and other data will be discussed (table 1).

Chemical Methods of Inactivation

There was little diminution in BSE infectivity after a 2-year exposure to formol saline [21]. This accords with the knowledge that other TSE agents resist inactivation by formalin and other aldehydes [22]. BSE infectivity was inactivated by exposure for 30 min to solutions of sodium hypochlorite containing 16,5000 ppm available chlorine [23]. In contrast, a sodium dichloroisocyanurate solution with an equivalent concentration of available chlorine was not effective [23]. Studies with BSE-infected bovine brain and scrapie-infected rodent brain showed that treatment with 1 or 2 M sodium hydroxide for up to 2 h did not completely inactivate these agents, and permitted the survival of up to 4 logs of infectivity [23]. This contradicts earlier data showing that a 1-hour treatment with 1 M sodium hydroxide was effective [24] but the sensitivity of these assays was substantially reduced by the need to dilute the samples before injection to render them non-toxic. In a more recent study, it was not found necessary to dilute the samples if the pH was carefully neutralized before injection, and the assays were therefore more sensitive [23]. Other reports record the detection of residual scrapie infectivity after treatment with 1 M sodium hydroxide for either 1 h [25, 26] or 24 h [27], and the survival of CJD infectivity after exposure to 1 M [28] or 2 M sodium hydroxide [29].

Heat Treatment

The 22A strain of scrapie agent was not inactivated by microwave irradiation [30]. Dry heat at temperatures up to 180 °C for 1 h did not inactivate the ME7 strain of scrapie agent, and there was some survival of infectivity after exposure at 160 °C for 24 h; a 1-hour treatment at 200 °C was effective [31]. However, a significant degree of survival of the 263K strain of hamster-passaged scrapie agent and the 301V strain of mouse-passaged BSE agent after a 1-h exposure at 200 °C has been demonstrated [32]. Other data have shown some survival of 263K infectivity after exposure to 360 °C for 1 h [33]. However, lyophilized brain homogenate was heated under anoxic conditions in this study; prior drying is known to enhance the thermostability of conventional micro-organisms and TSE agents [34, 35]. Undiluted macerates (350 mg) and saline homogenates of BSE-infected bovine brain were exposed to GD autoclaving at 132 °C. Survival of infectivity in both types of sample after a 30-min exposure was not surprising, given that CJD and scrapie agents had been shown previously to survive after a 30-min, but not 1-hour exposure [24]. After a 1-hour exposure, the BSE-infected macerate, but not the homogenate, was still infectious [36]. Others have reported some survival of scrapie infectivity after infected brain homogenates were exposed to GD autoclaving at 132 °C for 1 h [26, 37].

Previously described differences in the thermostability of mouse-passaged strains of scrapie agent [38] were confirmed in the studies of Kimberlin et al. [18]. Although strain 139A was completely inactivated by exposure to GD autoclaving at 126 °C for 2 h, strain 22A was not; a 4-hour exposure of 22A is required to inactivate 22A under these conditions [39]. However, the studies of Kimberlin et al. [18] also showed that PL autoclaving at 136 °C for 4 min was completely effective with both of these strains of scrapie agent, and resulted in the UK recommendation to use PL autoclaving at 134–138 °C for 18 min for inactivating CJD-contaminated materials [17]. Nevertheless, it was still recommended that instruments used in surgery involving the brain, spinal cord or eyes of known or suspected cases of CJD should be discarded rather than recycled after autoclaving. This advice was later extended to include other categories of patients recognized to have a higher risk of developing CJD. These are blood relatives of families with a known predisposition to TSE, and individuals who had been recipients of (a) hormones derived from the pituitary glands of human cadavers (b) dura mater graft material derived from human cadavers, or (c) human corneal grafts. The continuing advice not to recycle surgical instruments after their use in neurosurgical or ophthalmo-logical procedures was probably based upon the history of doubt about the effectiveness of autoclaving with TSE agents, and further studies did cast doubt on the reliability of the PL autoclaving standard. BSE agent and two strains of rodent-passaged scrapie agent survived exposure to such PL cycles even when the exposure period was increased to 1 h [23]. However, the average mass of the infected brain-macerates used in this study was 340 mg [23], compared with 50 mg in the earlier study [18]. The larger samples were used because similarly sized samples of intact (unmacerated) brain tissue had been previously inactivated by the 134–138 °C PL procedure [36, 40, 41]. It was also considered that the larger samples might more realistically represent the maximum mass of TSE-infected tissue that might have to be disposed of by autoclaving during human and veterinary healthcare, but no official advice has ever been issued in this respect. As will be discussed, the degree of smearing and drying onto the glass containers that occurred with the larger (340 mg) samples is the main explanation for the survival of infectivity in these samples. Therefore, partial drying of infected tissue onto glass or metal surfaces should be a prerequisite when trying to define effective standards for inactivating TSE agents by heat.

Because of the uncertainties relating to PL autoclaving introduced by the studies of Taylor et al. [23], further experiments were carried out to assess the effectiveness of PL autoclaving cycles at 134, 136 and 138 °C for times ranging between 9 and 60 min using samples of infected brain macerates weighing either 50 or 375 mg. The agents used were (a) 22A, a mouse-passaged strain of scrapie agent that is known to be more thermostable than other strains of mouse-passaged scrapie agent [18, 38], (b) 263K, a hamster-passaged strain of scrapie agent that had more recently been shown to survive PL autoclaving [23], and (c) 301V, a mouse-passaged strain BSE agent that had not been tested previously. The data from these experiments [42] show that 301V can survive exposure to 138 °C for 1 h. However, 50-mg macerates of 22A-infected brain-tissue in which the infectivity levels were $\geq 10^{7.2}$ ID_{50} were inactivated by all of the 136 °C processes, which accords with earlier data [18]; the same is true for the 50-mg macerates exposed for four different time periods at 134 °C. Paradoxically, 1 case was observed in mice injected with material from a 50-mg sample autoclaved at 138 °C for 9 min. This might have been written off as an experimental aberration, had it not been that positive cases were also observed in mice injected with material from 375-mg macerates autoclaved at 136 or 138 °C (but not at 134 °C). These data suggest that the thermostability of the 22A strain was actually enhanced as the temperature of autoclaving was increased, and the difference between the 134 and 138 °C samples was statistically significant ($p < 0.01$). With 263K the starting titre was $10^{8.3}$ ID_{50}/g, and there was much the same degree of survival of the agent whether autoclaving was carried out at 134, 136, or 138 °C which would support the above hypothesis. For 301V which had a starting titre of $10^{8.6}$ ID_{50}/g the data are even more convincing in this respect; 60% of the animals injected with material autoclaved at 134 °C developed disease; the ratio for samples exposed at 138 °C was 72%. This is statistically significant ($p < 0.05$). These data indicate that simply increasing PL autoclaving temperatures and holding times would not necessarily be effective in achieving a reliable decontamination standard for inactivating TSE agents.

Observations on the Thermostability of TSE Agents

When scrapie agent is completely inactivated by autoclaving, destruction of the agent proceeds in an exponential fashion [43]. If the amounts of infectivity remaining after increasing exposure times, through to the time when complete inactivation is achieved, are plotted on a logarithmic scale, a straight line is obtained which shows that the death rate is constant. In contrast, when a heating procedure is only partially inactivating, a tailing type of inactivation curve results which shows an initial decline and then flattens and persists with time [43]. After autoclaving at 134–138 °C for 18 min, it has been shown that the amount of BSE or scrapie infectivity that survives is relatively constant regardless of either the starting titre, or whether the agents are present in bovine, hamster or mouse brain [44]. An example of the tailing type of inactivation curve can be derived from data relating to the inactivation of the 263K strain of scrapie agent by autoclaving [23]. If, as is often the practice with conventional micro-organisms, the initial steep decline in infectivity is used to predict the time that it will take to achieve complete inactivation, this results in a gross underestimate for 263K. This reinforces the viewpoint that, although such estimates may sometimes by useful, there is no substitute for establishing full inactivation curves [45]. Tailing inactivation curves are not uncommon for conventional micro-organisms; these may result from the protective effect of aggregation during the inactivation process, or be due to population heterogeneity where differing straight-line inactivation curves for two or more subpopulations combine to produce a tailing curve. Where there is no population heterogeneity, the same sort of tailing curve is usually obtained when the surviving organisms are recultured and retested [46]. One explanation for the presence of heat- or chemical-resistant subpopulations of scrapie agent might be the protective effect of aggregation which could occur in homogenates of infected tissue but not in undiluted tissue. This argument has been invoked to explain why 2% peracetic acid inactivated the scrapie infectivity in intact scrapie-infected mouse brain but not in 10% homogenates of brain tissue [47].

Although 7 logs of infectivity were lost, 2 logs of 263K in 340-mg samples of macerated hamster brain survived autoclaving at 134 °C for 1 h. Similarly sized samples of infected brain tissue are completely decontaminated within 18 min by autoclaving if the brain tissue is undisrupted, as opposed to macerated. The lesser efficiency of inactivating macerates may result from the fact that some smearing and drying before autoclaving occur with this type of sample [34, 35]. It seems that PrP^{Sc} in the dried portions of the brain macerates is rapidly heat-fixed but retains its biological activity, despite the ensuing exposure to high-pressure steam. Protection by fixation has been shown to occur during the inactivation of poliovirus by formalin [50], and prior fixation in ethanol [36] or formalin [41] considerably enhances the thermostability of the scrapie agent. It has also been observed that the amount of scrapie infectivity inactivated after 4 h under vacuum at 72 °C (resulting in low-temperature boiling) is greater than that achieved using the same equipment over the same timescale at atmospheric pressure when an end temperature of 120 °C is achieved due to the presence of fat [16]. The relatively heat-resistant subpopulation of scrapie agent retains its thermostability when reheated, suggesting that this is an acquired but stable characteristic of the heat-resistant subpopulation that differentiates it from the main population. After one PL autoclaving cycle at 134 °C for 18 min, the titre of 263K was reduced by 3.3 logs, but only by a further 1.7 logs after autoclaving for a second time [48]. Evidence for the intrinsic and fundamental difference of this subpopulation comes from the fact that, at its limiting dilution, it produces an average incubation period which is well beyond that for unheated material at its limiting dilution [49]. Collectively, the data show that procedures which produce rapid and/or extremely effective fixation of PrP^{Sc} result in enhanced resistance of TSE agents to heat inactivation.

Combining Heat with Exposure to Sodium Hydroxide

Although autoclaving, or exposure to sodium hydroxide, do not per se completely inactivate TSE agents, inactivation can be achieved by combining these procedures. Taguchi et al. [51] and Ernst and Race [26] described the successful inactivation of CJD and scrapie infectivity respectively by exposure to 1 M sodium hydroxide followed by GD autoclaving at 121 °C for 30 or 60 min. Inactivation of 263K has also been reported after GD autoclaving at 121 °C for 90 min in the presence of 1 M sodium hydroxide [27]. More recently, it has been observed that if 22A is autoclaved at 121 °C for 30 min in the presence of 2 M sodium hydroxide (without a prior holding period in sodium hydroxide), inactivation can be achieved [52]. There are practical problems relating to this procedure, such as the potential exposure of operators to splashing with sodium hydroxide, and the potential deleterious effect on the autoclave. There are now data that show inactivation of the high-titre 301V strain of mouse-passaged BSE agent by boiling in 1 M sodium hydroxide for 1 min [53].

Current Concerns

At present, there is no way of determining the potential scale of the vCJD epidemic. In contrast to the situation with sporadic CJD, PrP^{Sc} has been detectable in lymphoreticular tissue such as tonsil, spleen and lymph nodes in vCJD [54]. In sheep with scrapie, it is known that such tissues become infected long before the onset of clinical signs. The same is true for scrapie in mice, and is likely to be the case with vCJD, as evidenced by the detection of PrP^{Sc} in the appendix of a pre-clinical case of vCJD [55]. Depending upon the number of individuals in the UK currently incubating vCJD, instruments used in general surgery (not just neurosurgery) could increasingly pose a risk with regard to iatrogenic transmission, and it is intended that archived samples of appropriate tissues should be screened for PrP^{Sc} to quantify this. If there is a significantly increasing risk, it may be that processing surgical instruments through hot sodium hydroxide could be an effective option, given that this is a highly inactivating process. However, this would require the identification of types of stainless steel that would not be adversely affected. A non-damaging process that is under investigation is autoclaving at 121 °C in 5% sodium dodecyl sulphate. Although a 15-min exposure was not completely effective, there was a significant reduction in the infectivity titre [55], suggesting that a 30-min exposure might prove to be completely effective. There is clearly a need to establish reliable methods for decontaminating CJD-infected surgical instruments without damaging them.

References

1   Prusiner SB: Molecular biology of prion diseases. Science 1991;1252:1515–1522.
2   Prusiner SB: Prion diseases and the BSE crisis. Science 1997;278:245–251.
3   Almond J, Pattison J: Human BSE. Nature 1997;389:437–438.
4   Coles H: Nobel panel rewards prion theory after years of heated debate. Nature 1997;389:529.
5   Chesebro B: BSE and prions: Uncertainties about the agent. Science 1998;279:42–43.
6   Farquhar CF, Somerville RA, Bruce ME: Straining the prion hypothesis. Nature 1998;391:345–346.
7   Bruce ME, Chree A, McConnell I, Foster J, Pearson G, Fraser H: Transmission of bovine spongiform encephalopathy and scrapie to mice; strain variation and the species barrier. Philos Trans R Soc B 1994;343:405–411.
8   Bruce ME, Will RG, Ironside JW, McConnell I, Drummond D, Suttie A, McCardle L, Chree A, Hope J, Birkett C, Cousens S, Fraser H, Bostock CJ: Transmissions to mice indicate that 'new variant' CJD is caused by the BSE agent. Lancet 1997;389:498–501.
9   Will RG, Ironside JW, Zeidler M, Cousens SN, Estibeiro K, Alperovitch A, Poser S, Pocchiari M, Hofman A, Smith PG: A new variant from Creutzfeldt-Jakob disease in the UK. Lancet 1996;347:921–925.
10 Bernoulli C, Siegfried J, Baumgartner G, Regli F, Rabinowicz T, Gajdusek DC, Gibbs CJ: Danger of accidental person-to-person transmission of Creutzfeldt Jakob disease by surgery. Lancet 1997; i:478–479.
11 Foncin JF, Gaches J, Cathala F, El Sherif E, Le Beau E: Transmission iatrogène interhumaine possible de maladie de Creutzfeldt-Jakob avec atteinte des grains du cervelet. Rev Neurol 1980;136:280.
12 Greig JR: Scrapie in sheep. J Comp Pathol 1950;60:263–266.
13 Agrimi U, Ru G, Cardone F, Pocchiari M, Caramelli M: Epidemic of transmissible spongiform encephalopathy in sheep and goats in Italy. Lancet 1999;353:560–561.
14 Wilesmith JW, Wells GAJ, Cranwell MP, Ryan JBM: Bovine spongiform encephalopathy: Epidemiological studies. Vet Rec 1998;123:638–644.
15 Taylor DM, Woodgate SL, Atkinson MJ: Inactivation of the bovine spongiform encephalopathy agent by rendering procedures. Vet Rec 1995;137:605–610.
16 Taylor DM, Woodgate SL, Fleetwood AJ, Cawthorne RJG: The effect of rendering procedures on scrapie agent. Rec 1997;141:643–649.
17 DHSS: Management of patients with spongiform encephalopathy Creutzfeldt-Jakob disease (CJD). DHSS Circular DA (84) 16.

18 Kimberlin RH, Walker CA, Millson GC, Taylor DM, Robertson PA, Tomlinson AH, Dickinson AG: Disinfection studies with two strains of mouse passaged scrapie agent. J Neurol Sci 1983;59: 355–369.

19 Rosenberg RN, White CL, Brown P, Gajdusek DC, Volpe JJ, Posner J, Dyck P: Precautions in handling tissues, fluids, and other contaminated materials from patients with documented or suspected Creutzfeldt-Jakob disease. Ann Neurol 1986;19:75–77.

20 Federal Ministry of Health: Guidelines on safety measures in connection with medicinal products containing body materials obtained from cattle, sheep or goats for minimizing the risk of transmission of BSE or scrapie. Fed Bull 1994;40:February.

21 Faser H, Bruce ME, Chree A, McConnell I, Wells GAH: Transmission of bovine spongiform encephalopathy and scrapie to mice. J Gen Virol 1992;173:1891–1897.

22 Taylor DM: Transmissible degenerative encephalopathies: Inactivation of the unconventional transmissible agents; in Russell AD, Hugo WB, Ayliffe GAJ (eds): Principles and Practice of Disinfection, Preservation and Sterilization. London, Blackwell, 1999, pp 222–236.

23 Taylor DM, Fraser H, McConnell I, Brown DA, Brown KL, Lamza KA Smith GRA: Decontamination studies with the agents of bovine spongiform encephalopathy and scrapie. Arch Virol 1994; 139:313–326.

24 Brown P, Rohwer RG, Gajdusek DC: Newer data on the inactivation of scrapie virus or Creutzfeldt-Jakob disease virus in brain tissue. J Infect Dis 1986;153:1145–1148.

25 Diringer H, Braig HR: Infectivity of unconventional viruses in dura mater. Lancet 1989;i:439–440.

26 Ernst DR, Race RE: Comparative analysis of scrapie agent inactivation. J Virol Methods 1993;41: 193–202.

27 Prusiner SB, McKinley MP, Bolton DC, Bowman KA, Groth DF, Cochran SP, Hennessey EM, Braunfeld MB, Baringer JR, Chatigny MA: Prions: Methods for assay, purification, and characterisation. Methods Virol 1984;8:293–345.

28 Tamai Y, Taguchi F, Miura S: Inactivation of the Creutzfeldt-Jakob disease agent. Ann Neurol 1988;24:466–467.

29 Tateishi J, Tashima T, Kitamoto T: Inactivation of the Creutzfeldt-Jakob disease agent. Ann Neurol 1988;24:466.

30 Taylor DM, Diprose MF: The response of the 22A strain of scrapie agent to microwave irradiation compared with boiling. Neuropathol Appl Neurobiol 1996;22:256–258.

31 Taylor DM, McConnell I, Fernie K: The effect of dry heat on the ME7 strain of scrapie agent. J Gen Virol 1996;77:3161–3164.

32 Steele PJ, Taylor DM, Fernie K: Survival of BSE and scrapie agents at 200°C. Abstracts of a Meeting of the Association of Veterinary Teachers and Research Workers, Scarborough, March 1999, p 21.

33 Brown P, Liberski PP, Wolff A, Gajdusek DC: Resistance of scrapie agent to steam autoclaving after formaldehyde fixation and limited survival after ashing at 360 °C: Practical and theoretical implication. J Infect Dis 1990;161:467–472.

34 Asher DM, Pomeroy KL, Murphy L, Rohwer RG, Gibbs CJ, Gajdusek DC: Practical inactivation of scrapie agent on surfaces. Abstracts of the IXth International Congress of Infectious and Parasitic Diseases, Munich, July 1986.

35 Asher DM, Pomeroy KL, Murphy L, Gibbs CJ, Gajdusek DC: Attempts to disinfect surfaces contaminated with etiological agents of the spongiform encephalopathies. Abstracts of the VIIth International Congress of Virology 1987, Edmonton, August 1987, p 147.

36 Taylor DM: Transmissible subacute spongiform encephalopathies: Practical aspects of agent inactivation; in Court L, Dodet D (eds): Transmissible Subacute Spongiform Encephalopathies: Prion Disease. IIIrd International Symposium on Subacute Spongiform Encephalopathies: Prion Diseases, Paris, March 1996, pp 479–482.

37 Pocchiari M: Unpublished data cited by Horaud F: Safety of medicinal products: Summary. Dev Biol Stand 1993;80:207–208.

38 Dickinson AG, Taylor DM: Resistance of scrapie agent to decontamination. N Engl J Med 1978; 229:1413–1414.

39 Dickinson AG: Scrapie in sheep and goats; in Kimberlin RH (ed): Slow Virus Diseases of Animals and Man. Amsterdam, North-Holland, 1976, pp 209–241.

40   Taylor DM: Decontamination of Creutzfeldt-Jakob disease agent. Ann Neurol 1986;20:749.
41   Taylor DM, McConnell I: Autoclaving does not decontaminate formol-fixed scrapie tissues. Lancet 1988;i:1463–1464.
42   Taylor DM: Inactivation of prions by physical and chemical means. J Hosp Infect 1999;43(suppl): S69–S76.
43   Rohwer RG: Scrapie inactivation kinetics – An explanation for scrapie's apparent resistance to inactivation – A re-evaluation of estimates of its small size; in Court LA, Cathala F (eds): Virus nonconventionnels et affections du système nerveux central. Paris, Masson, 1983, pp 84–113.
44   Taylor DM: Creutzfeldt-Jakob disease. Lancet 1996;347:1333.
45   Greene VW: Sterility assurance: Concepts, methods and problems; in Russell AD, Hugo WB, Ayliffe GAJ (eds): Principles and Practice of Disinfection, Preservation and Sterilization. Oxford, Blackwell, 1992, pp 605–624.
46   Gardner JF, Peel MM: Introduction to Sterilization, Disinfection and Infection Control. Edinburgh, Churchill Livingstone, 1991.
47   Taylor DM: Resistance of the ME7 scrapie agent to peracetic acid. Vet Microbiol 1991;27: 19–24.
48   Taylor DM, Fernie K, McConnell I, Steele P: Observations on thermostable subpopulations of the unconventional agents that cause transmissible degenerative encephalopathies. Vet Microbiol 1998; 64:33–38.
49   Taylor DM, Fernie K: Exposure to autoclaving or sodium hydroxide extends the dose-response curve of the 263K strain of scrapie agent in hamsters. J Gen Virol 1996;77:811–813.
50   Gard S, Maaloe O: Inactivation of viruses; in Burnet FM, Stanley WM (eds): The Viruses. New York, Academic Press, 1959, vol 1, pp 359–427.
51   Taguchi F, Tamai Y, Uchida K, Kitajima R, Kojima H, Kawaguchi T, Ohtani Y, Miura S: Proposal for a procedure for complete inactivation of the Creutzfeldt-Jakob disease agent. Arch Virol 1991; 119:297–301.
52   Taylor DM, Fernie K, McConnell I: Inactivation of the 22A strain of scrapie agent by autoclaving in sodium hydroxide. Vet Microbiol 1997;58:87–91.
53   Taylor DM, Fernie K, Steele P: Boiling in sodium hydroxide inactivates mouse-passaged BSE agent. Abstracts of a Meeting of the Association of Veterinary Teachers and Research Workers, Scarborough, March 1999, p 22.
54   Hill AF, Butterworth RJ, Joiner S, Jackson G, Rossor MN, Thomas DJ, Frosh A, Tolley N, Bell JE, Spencer M, King A, Al-Sarraj S, Ironside JW, Lantos PL, Collinge J: Investigation of variant Creutzfeldt-Jakob disease and other human prion diseases with tonsil biopsy samples. Lancet 1999; 353:183–189.
55   Hilton DA, Fathers E, Edwards P, Ironside JW, Zajicek J: Prion immunoreactivity in appendix before clinical onset of variant Creutzfeldt-Jakob disease. Lancet 1998;352:703–704.
56   Taylor DM, Fernie K, McConnell I, Steel PJ: Survival of scrapie agent after exposure to sodium dodecyl sulphate and heat. Vet Microbiol 1999;67:13–16.

D.M. Taylor, Neuropathogenesis Unit, Institute for Animal Health,
West Mains Road, Edinburgh EH9 3JF (UK)
Tel. +44 131 667 5204, Fax +44 131 668 3872, E-Mail david-m.taylor@bbsrc.ac.uk

Rabenau HF, Cinatl J, Doerr HW (eds): Prions. A Challenge for Science,
Medicine and Public Health System. Contrib Microb. Basel, Karger, 2001, vol 7, pp 68–92

....

Human Prion Diseases: Cause, Clinical and Diagnostic Aspects

Richard Knight [a], *Steven Collins* [b]

[a] National CJD Surveillance Unit, Western General Hospital, Edinburgh, UK, and
[b] National CJD Registry, Department of Pathology, The University of Melbourne,
Parkville, Victoria, Australia

There are four predominant human prion disease phenotypes: Creutzfeldt-Jakob disease (CJD); Gerstmann-Sträussler-Scheinker syndrome (GSS); fatal familial insomnia (FFI), and kuru. The commonest is CJD, varying in causation from genetic through acquired to unknown, with different epidemiological and clinical characteristics (table 1). However, all prionoses are progressive, invariably fatal neurodegenerative diseases, with broadly similar neuropathological features, most specifically in terms of PrP^{Sc} deposition, as well as their transmissibility to a range of laboratory animal hosts.

Our understanding of these diseases, especially that which has arisen from molecular biology, suggests a need for nosological reconsideration. CJD has become a generic diagnostic label for syndromes with different aetiologies (idiopathic, acquired and inherited) and with rather different clinico-pathological phenotypes (as is the case, for example, with sporadic and variant CJD). GSS, originally delineated as an autosomal dominantly inherited cerebellar ataxia with a characteristic pathology, is now something of an umbrella term, covering a group of illnesses with differing clinico-pathological features and varied underlying genetic abnormalities. In contrast, typical FFI remains a distinctive clinico-pathological entity linked to a specific genetic mutation. However, the same mutation may give rise to a clinico-pathological picture of CJD or FFI, depending on the background genotype at codon 129 of the prion protein gene (*PRNP*). Hence, GSS and CJD are perhaps redundant historical terms for forms of genetic human prion disease, with the term 'CJD' possibly regarded as somewhat anachronistic. Certainly, the different subtypes have been confused, a misleading uniformity arising from the single eponymous

Table 1. Types of human prion disease

Disease	Distribution	Cause	Notes
CJD	worldwide	unknown	1:1000,000/year
GSS	familial	genetic	rare
FFI	familial	genetic	extremely rare
Kuru	Papua New Guinea	ritual cannibalism	vanishing

label. Kuru is perhaps the least ambiguous prion disease, because of its extraordinary cause and epidemiology. The clinical features of prion diseases in general, depend on the agent strain, the specific cause or mode of exposure and the genotype of the affected individual.

Creutzfeldt-Jakob Disease

CJD is the most significant of the human prion diseases, partly because it is the commonest, but also because it includes variant CJD (vCJD) which is of such current medical, economic and political importance. The first description of CJD was published in 1921 by Dr. A. Jakob. Most modern commentators regard the inclusion of Dr. H. Creutzfeldt's 1920 paper as erroneous, but this has not eroded the use of the oddly compelling double eponym [1]. Perhaps inevitably, the original description was followed by a number of papers describing clinical and pathological variations, often given their own eponymous names (such as Heidenhain's syndrome) [2]. In 1968, laboratory transmission to the chimpanzee was reported by Gibbs et al. [3], prompting seminal unification of the disease subtypes and also raising the question as to whether this rare, apparently neurodegenerative illness, might be some peculiar sort of natural infection. The clinical and epidemiological studies that have followed this unexpected report have led to recognition of four main forms of CJD: sporadic, iatrogenic, variant and genetic (table 2).

Sporadic Creutzfeldt-Jakob Disease

Epidemiology and Cause
This is the commonest form of CJD, but with a worldwide incidence of only approximately 1 per million of the population per year. Recent published incidence rates have tended to be higher than those reported in the 1970–1980s,

Table 2. Types of CJD

CJD type	Distribution	Cause
Sporadic	worldwide	unknown
Iatrogenic	reported in several countries	accidental medical transmission
New variant	UK, France, Republic of Ireland	probable BSE-contaminated food
Genetic	worldwide with particular foci	genetic

but this probably reflects better case ascertainment and reporting consequent to current interest [4]. Overall, males and females are affected approximately equally, although some studies have reported a slight female excess [5]. Sporadic CJD (spCJD) is predominantly a disease of late middle age with a mean age at death in the late 60s (fig. 1). Cases with onset below the age of 50 are relatively rare. However, the incidence does not rise consistently with age and the often reported decline in incidence after the age of 70 years has been the subject of some debate. Under-ascertainment in the elderly is one possible explanation, perhaps supported by the finding of a particular rise in elderly incidence rates in recent UK surveillance reports [4]. A number of studies in different countries have examined the incidence and characteristics of spCJD and have conducted case-control studies of potential risk factors [6]. The cause of spCJD is unknown. Research has failed to find any link with scrapie, either by country, diet or occupation. There is no convincing or consistent evidence of significant clustering of cases over time, and direct case-to-case contact is not a likely mode of transmission. The possibility of unsuspected iatrogenic transmission via contamination of surgical instruments or blood transfusion is not supported by most surveillance data nor by case-control studies, although a recent Australian publication reported an excess of surgical operations in cases compared to controls [7]. Alternative explanations include a spontaneous conformational change of the normal prion protein (PrP^C) to PrP^{Sc} (the relatively protease-resistant, disease-associated isoform) or a somatic mutation of the *PRNP* gene.

In the normal human population, there is a common polymorphism at codon 129 of the *PRNP* gene open reading frame (situated on chromosome 20), whereby coding may be for methionine or valine, with individuals being MM or VV homozygotes or MV heterozygotes. In the normal UK population, the distributions are approximately: MM 40%; VV 10%, and MV 50%. However, approximately 80% of UK spCJD cases harbour the MM genotype, with only about 10% being the MV genotype. This indicates that being an MM homozygote is, in some way, a risk factor for spCJD and that MV heterozygotes

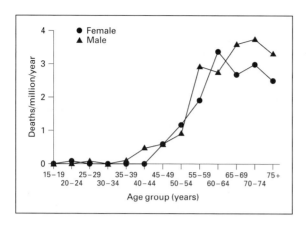

Fig. 1. Age- and sex-specific mortality rates from spCJD in the UK (1995–1999). Mortality rates were calculated using 1991 census.

are relatively protected. This genetic factor has been found in other studied populations [5].

Clinical Features

Various prodromal symptoms have been described (including, headache, tiredness, sleep or appetite disturbance, and depression), but these are entirely non-specific, perhaps being given retrospective significance because of later developments. A case-control study in England and Wales between 1980 and 1984, using hospital patient controls, did not suggest that any of these symptoms were specifically related to CJD [8]. The presenting neurological symptoms often reflect a relatively focal neurological involvement but, in most cases, they are soon subsumed in a rapidly progressive global encephalopathy. Cognitive impairment and cerebellar ataxia are the commonest early features. Two particularly well-recognized, but relatively rare, presentations are the Heidenhain variant, with cortical blindness, and the Brownell-Oppenheimer variant, with a pure cerebellar syndrome [2, 9]. Whatever the mode of onset, the subsequent clinical course in most cases, is one of rapid progression, often surprising the clinician and dismaying the relatives. The rapidly evolving clinical picture always includes dementia, with ataxia and myoclonus being present in the majority of cases. Other features include primitive reflexes, pyramidal signs, cortical blindness and paratonic rigidity. The terminal state is often one of akinetic mutism. Major pyramidal weakness, unequivocal lower motor neuron features and epileptic seizures are relatively uncommon. The median duration of spCJD is 4 months and around 65% of cases have an

Table 3. Diagnostic criteria for spCJD

I		Neuropathological (including immunocytochemical) confirmation
II		Rapidly progressive dementia
III	A	Visual/cerebellar features
	B	Pyramidal/extrapyramidal features
	C	Myoclonus
	E	Akinetic mutism
IV		Typical periodic EEG
V		Positive CSF 14-3-3

Definite:	I
Probable:	II + 2/4 of III + IV or possible + V
Possible:	II + 2/4 and duration < 2 years

illness duration of less than 6 months. In 14% of cases, there is a relatively long duration of 12 months or more. However, durations of greater than 2 years are rare (5%) and this fact is reflected in the time limit embodied in the clinical diagnostic criteria for spCJD (discussed below).

The polymorphism at codon 129 of *PRNP* not only affects susceptibility but also influences the clinico-pathological expression of disease; the clinical and neuropathological features of spCJD varying somewhat according to whether the individual has the codon 129 MM, VV or MV genotype [10].

Diagnosis

In a typical case, the invariably progressive, rapid clinical evolution and the short illness duration, readily distinguish spCJD from most other dementing illnesses. However, the differential diagnosis can be problematic, especially if the onset of disease is relatively abrupt, the initial symptoms (such as ataxia) remain focal for a longer period, or there is an unusually long illness duration. In one UK study, about 6% of cases were initially misdiagnosed as stroke [11]. Alzheimer's disease with a less typical, relatively rapid myoclonic progression may occasionally cause confusion. As with many illnesses, a careful history is vital. The diagnosis of spCJD is presently categorized as definite, probable and possible (table 3). Definite spCJD requires neuropathological confirmation, usually obtained at autopsy but, occasionally, cerebral biopsy might be considered. Probable and possible diagnoses are clinical, and are based on the exclusion of alternative diagnoses, with the presence of requisite characteristic features and supportive findings in the EEG or CSF. A probable classification carries an approximately 95% certainty of spCJD and this reliability of the clinical criteria is one reason why biopsy confirmation is seldom necessary. However, it should be stressed that there is no infallible pathognomonic clinical diagnostic test, autopsy being necessary for diagnostic certainty. The clinical diagnostic criteria need careful application and the pre-mortem assessment of suspect cases requires neurological expertise. The EEG shows characteristic periodic discharges at some stage of the illness in about two thirds of cases, but the abnormality may not be present until late in the illness and repeat recordings (usually at weekly intervals) may be necessary. The absence of this EEG pattern cannot exclude the diagnosis and the typical discharges are seen in conditions other than CJD, occasionally even in Alzheimer's disease. A persistently normal EEG, despite clinical progression, is incompatible with a diagnosis of spCJD [12]. The CSF 14-3-3 test is a valuable supportive investigation. 14-3-3 is a normal brain protein that may be present in the CSF in excessive amounts following neuronal injury. It is not intrinsically associated with CJD, and, in particular, is not specifically related to PrP. 14-3-3 may be elevated in a number of conditions including viral encephalitis, subarachnoid haemorrhage and recent cerebral infarction. It is therefore perhaps surprising to find that, in the appropriate clinical context, it has a high degree of sensitivity (94%) and specificity (94%) for spCJD [13]. This specificity is largely because the other conditions which characteristically lead to a positive 14-3-3 CSF test are relatively easily distinguished from CJD on other grounds. Aside from an elevation in the total protein level, no other CSF abnormality is found and, in particular, pleocytosis is not a typical feature of CJD. Cerebral imaging is important in the investigation of suspect cases of spCJD, largely in order to exclude alternative diagnoses, and cerebral CT is characteristically normal or shows non-specific atrophy. Cerebral MRI is additionally important because, in some cases, a characteristic signal change is seen in the putamen and caudate. Occasionally, high signal is seen in the cerebral cortex and this may be focal, reflecting the particular clinical features of the case being investigated [14]. Significant atrophy is unusual if imaging is undertaken within 3 months of illness onset. Indeed, one early diagnostic clue is the notable finding of normal brain imaging in the context of a rapidly progressive and devastating cerebral illness. Routine blood tests do not show any relevant abnormality. Although modest disturbances in liver function tests may be seen in a number of cases, these probably reflect disturbances of general health consequent on the dementing illness or the effects of any symptomatic drug treatment.

Cerebral biopsy should be considered only if there is a reasonable possibility of an alternative, potentially treatable disease. There are published guidelines concerning precautions to be taken during procedures involving potential CJD material [15].

Iatrogenic Creutzfeldt-Jakob Disease

Iatrogenic CJD (iCJD) is the accidental transmission of CJD during medical or surgical treatment.

Epidemiology and Cause

iCJD has resulted from neurosurgery (including the use of EEG depth electrodes), corneal grafting, human dura mater implants or exposure, and the use of human cadaveric growth hormone (hGH) and human pituitary gonadotrophin [16].

There are only 3 reported corneal-grafting cases and 4 reported human gonadotrophin cases [17–19]. Numerically, the most important relate to cadaveric-derived hGH treatment and the surgical use of human dura mater grafts.

Some epidemiological features of iCJD tend to reflect the cause. For example, hGH cases tend to be relatively young. These cases resulted from the use of cadaveric-derived hGH manufactured from pituitary glands harvested at autopsy, with some glands presumably being obtained from CJD cases, with processing insufficient to eradicate infectivity. The hormone was derived from pooled harvested glands with many thousands of glands in a single production lot. The treatments were administered over a period of time, with repeated injections increasing the risk of exposure to an infected batch. Over 100 hGH cases have been described in a number of countries including France, the UK, New Zealand, the Netherlands and the USA. The ratio of cases to numbers treated varies with country, being 1:50 in France, 1:100 in the UK and 1:500 in the USA [20]. Since 1985, cadaveric hGH has been withdrawn from use in most countries. The incubation period of European hGH-related disease ranges from 4.5 to over 25 years (mean 12 years), with an unexplained longer period in USA cases (mean of nearly 20 years). The incubation periods are calculated from the mid-point of therapy to the onset of CJD symptoms.

Human cadaveric dura mater has been used in surgical and interventional procedures, particularly in neurosurgical operations to repair dural defects. Over 100 dura-mater-related CJD cases have been reported, from a number of countries, with 65 of these being identified in Japan alone [20]. Nearly all have been associated with a particular brand, namely Lyodura and the biggest risk seems to be associated with grafts used between 1981 and 1987. Human dura mater grafts are no longer used in many countries, but they are still used in the USA, with donor screening and special processing precautions. The incubation period for dura mater cases seems to vary with the site of grafting from 1–16 years.

Clinical Features

The clinical features depend somewhat on the route of infection. hGH cases resulting from the intra-muscular injection of infected material, generally present with a progressive cerebellar syndrome and other features including dementia tend to occur later [16]. Human dura mater cases, resulting from infected material being placed relatively near to the CNS tend to present with a rapidly progressive dementia along with other neurological features and may be clinically indistinguishable from sporadic cases, aside, of course, from the relevant past medical history.

Diagnosis

The principal indication of iatrogenic disease must come from the history, in particular a history of treatment with cadaveric-derived hGH or an operation involving the use of a human dura mater graft.

Variant Creutzfeldt-Jakob Disease

Much of the current interest in human prion diseases stems from the identification of this relatively new disease in 1996 [21]. Originally termed 'new variant CJD', it is now designated simply as 'variant CJD' (vCJD) although the organization set up by relatives of affected individuals in the UK has preferred the term 'human BSE'.

Epidemiology and Cause

Perhaps the most striking epidemiological feature of vCJD is its geographical distribution. In June 2000, 64 cases had been confirmed worldwide with only 3 outside the UK; 1 case being in the Republic of Ireland and 2 in France. The relatively young age of the affected individuals is also notable. In stark contrast to the age distribution of spCJD, the median age of onset in vCJD is 28 years (range 14–53). Males and females are affected equally. There is now convincing evidence that vCJD is due to infection of humans with the BSE agent, most probably via contaminated food. Experimental studies have demonstrated that the same agent strain causes both vCJD and BSE. The available epidemiological evidence strongly indicates that the agent passed from cattle to man and it is obviously relevant that the UK had the greatest incidence of BSE. To date, it has not been possible to provide absolute proof that the zoonotic link was via food, but this is a very reasonable hypothesis, and the most likely route is via mechanically recovered meat (MRM) included in certain prepared products. During the relevant period, some MRM is likely to have contained spinal cord material from infected cattle. Other means of transmission might be relevant, such as occupational contact, surgical sutures derived from bovine gut, and drugs or vaccines prepared using fetal calf serum, but no specific evidence currently exists to implicate these routes [22].

As with spCJD, the genotype of the individual appears to be important. All the cases tested to date have been codon 129 MM homozygotes, indicating that this particular genotype confers particular susceptibility to the disease. It also suggests that continued surveillance needs to be conducted with care: if BSE affects humans with MV or VV genotypes, the resulting clinical and pathological picture may be different.

It is still not known how many cases will eventually occur, as predictions must be based on rather uncertain assumptions about the population exposed to infection, the susceptibility of humans to the infection, and the resulting range of incubation periods. The decline in BSE along with the relevant precautions and regulations in place should prevent further instances of transmissions from cattle to man; however, there is concern that secondary human-to-human spread may occur. There is evidence of lymphoreticular involvement in vCJD (spleen, appendix, tonsil and lymph node contain PrP^{Sc}, which is not seen in other forms of CJD) and this raises the possibility of infection via contamination of surgical instruments and even via blood transfusion or the use of blood products [23].

Clinical Features

Whereas spCJD presents with rapidly progressive symptoms that are usually clearly neurological in nature, vCJD tends to present with difficulties that are not unambiguously neurological and with a relatively slower progression. The commonest presentation is that of a psychiatric/behavioural disturbance and/or unpleasant sensory symptoms including pain, usually without neurological abnormality on examination [24–26]. Many have seen a psychiatrist at some stage of the illness, some even being initially referred to psychiatry services by their general practitioner. In the majority of cases, features of depression are present and many were given antidepressant treatment [24, 25]. Other features included anxiety, agitation, delusions and hallucinations.

The sensory features are usually persistent, unpleasant or frankly painful, and may be worse at night. There is usually no definite abnormality on clinical sensory examination, although relevant signs are found occasionally. In some individuals, limb pain alone resulted in investigations such as an x-ray of relevant bones or joints. Nerve conduction studies were undertaken in some, but were normal. Where somatosensory-evoked potentials were performed, minor abnormalities were seen, suggesting a relatively central disturbance of sensory pathways. In some cases there was lateralization, which might suggest a cerebral origin. It is possible that the sensory symptoms reflect a thalamic

Table 4. Key neurological features in vCJD (n = 35)

Clinical feature	Cases affected	With feature from onset
Sensory symptoms	24	7
Limb pain	13/24	4/7
Ataxia	35	3
Involuntary movements	33[1]	2
Dystonia	12	2
Chorea	20	0
Myoclonus	25	0
Upgaze paresis	14	0
Dementia	35	0

[1] Two cases excluded because of insufficient clinical information.

disturbance. At some point, other neurological features develop and these are listed in table 4. Definite neurological abnormality develops at a mean of around 6 months from the first symptoms, with ataxia often being the most prominent problem. Eventually, the clinical picture is one of a dementia with multiple neurological features including myoclonus, although pre-terminal akinetic mutism is not as common as it is in spCJD. The illness duration is greater than that for spCJD, with a median of 14 months (range 6–38).

Diagnosis

The differential diagnosis of a progressive neuropsychiatric disorder in relative youth is arguably wider than that of a rapidly progressive dementia in middle to late life. In addition, there is now considerable accumulated experience with spCJD, and the clinical diagnostic criteria have been refined and validated over a number of years. Experience with vCJD is obviously relatively limited both in terms of numbers and time. The two investigations, namely the EEG and CSF 14-3-3, which have proved so helpful in the diagnosis of spCJD, are less so in vCJD. The typical periodic EEG pattern has never been seen in vCJD and the 14-3-3 test is less sensitive (and possibly less specific) than in sporadic disease. Fortunately, the cerebral MRI, which has had a rather limited role in spCJD, appears to be very useful in vCJD, with bilateral pulvinar high signal being seen in the majority of cases [27]. Clinical diagnostic criteria have been developed, including the MRI features, and are being pro-spectively evaluated in continuing UK surveillance (table 5) [28].

Table 5. Clinical diagnostic criteria FOR vCJD

I	A	Progressive neuropsychiatric disorder
	B	Duration of illness >6 months
	C	Routine investigations do not suggest an alternative diagnosis
	D	No history of potential iatrogenic exposure
II	A	Early psychiatric symptoms[1]
	B	Persistent painful sensory symptoms[2]
	C	Ataxia
	D	Myoclonus or chorea or dystonia
	E	Dementia
III	A	EEG does not show the typical appearance of spCJD[3] (or no EEG performed)
	B	Bilateral pulvinar high signal on MRI scan

Definite:	IA (progressive neuropsychiatric disorder) and neuropathological confirmation of nvCJD[4]
Probable:	I and 4/5 of II and IIIA and IIIB
Possible:	I and 4/5 of II and IIIA

[1] Depression, anxiety, apathy, withdrawal, delusions.
[2] This includes both frank pain and/or unpleasant dysaesthesia.
[3] Generalised triphasic periodic complexes at approximately one per second.
[4] Spongiform change and extensive PrP deposition with florid plaques, throughout the cerebrum and cerebellum.

Cerebral biopsy may be considered in selected cases. The main indication for this being significant concern about an alternative, potentially treatable, neurological illness.

Tonsillar biopsy has been proposed as a diagnostic test, but experience is very limited and this does not provide positive information about any possible alternative brain pathology [29]. With increasing experience of vCJD and the proposed clinical criteria, tissue diagnosis may not be necessary in life for the vast majority of cases.

Genetic Creutzfeld-Jakob Disease

Familial instances of CJD have been recognized since 1930, and GSS was first described around the same time. Familial CJD, and GSS and FFI, are almost always related to an underlying *PRNP* mutation and should perhaps be grouped together, but are considered separately below. The transmissibility of genetic CJD (gCJD) to non-human primates was first reported in 1973.

Table 6. Haplotypes linked with genetic prion disease

E200K,129M	D178N,129V[1]	D178N,129M[1,2]	A117V,129V[3]
V180I,129M	M232R,129M	P102L,129M[3]	P105L,129V[3]
T183A,129M	P102L,129M,219K	F198S,129V[3]	Q217R,129V[3]
H208R,129M	V210I,129M	Y145stop,129M[3]	Q212P[3,4]
P102L,129V			
ins24bp,129M	ins48bp,129M	ins96bp,129M	ins96bp,129V
ins120bp,129M	ins144bp,129M	ins168bp,129M	ins216bp,129M
ins192bp,129V[3]			

[1] Codon 129 type determines phenotype resulting from mutation.
[2] Linked to GSS phenotype.
[3] Linked to FFI phenotype.
[4] Codon 129 not specified.

Epidemiology and Cause

In gCJD, there is an underlying mutation of the *PRNP* gene open reading frame leading to the production of a PrP protein molecule theorised to have a heightened predisposition to adopting the configuration of PrP[Sc]. Hence, the mutation is believed to be directly causative of a familial disease with an autosomal dominant pattern of inheritance. However, the mutation is conceivably a susceptibility factor rendering the affected individual particularly liable to the effects of some, as of yet, unidentified environmental agent. A number of different (point and insert) mutations have been identified (table 6).

Genetic CJD tends to occur in clusters reflecting its familial nature. There are two very notable geographical concentrations of genetic disease, involving the E200K mutation. The first is in Israel, affecting Jews of Libyan origin, with an incidence of approximately 50 per million per year in this community. The second concerns two foci in Slovakia, with incidences ranging from around 4 to 17 per million per year.

Clinical Features

The clinical presentation varies with the underlying mutation and other factors (such as codon 129 genotype, which may affect the clinico-pathological phenotype resulting from any given mutation). For the Libyan Jewish E200K, cis-129M focus, the picture is not dissimilar to that of spCJD with a mean age at onset of 62 years, a mean disease duration of under a year, and a broadly similar clinical profile. However, a sensorimotor polyneuropathy is a typical additional clinical feature. The codon 129 characteristic of the normal allele does not appear to affect the disease phenotype. This is not so in D178N-

129V disease. Homozygous VV individuals have a mean age at onset of around 39 years with a mean illness duration of around 14 months, whereas the corresponding figures for heterozygous MV persons are 49 years and 27 months, respectively. The disease phenotype associated with insert mutations is particularly variable and may, in some instances, involve young age at onset (for example, in the third decade) and very long duration illnesses (even exceeding 10 years). At least one determining factor is the number of repeats in the particular abnormal allele.

Diagnosis

The firm diagnosis of genetic CJD requires pathological confirmation of CJD in an individual proven to harbour an underlying *PRNP* mutation. DNA can be extracted from a blood sample and the open reading frame of the gene sequenced. A family history of the disease with an autosomal pattern of inheritance is the usual background, but apparently sporadic cases of CJD occasionally prove to have an underlying mutation, and *PRNP* gene sequencing is the only certain means of definitively excluding genetic disease.

Gerstmann-Sträussler-Scheinker Syndrome

GSS is almost invariably genetically determined, associated with an underlying mutation in the *PRNP* gene open reading frame. Although this syndrome was intermittently reported under a variety of titles following its original description in a large Austrian family spanning a number of generations [30, 31], its nosological clarification as a prion disease was only achieved relatively recently [32]. Originally defined in terms of an autosomal dominantly inherited illness manifested predominantly as a progressive cerebellar ataxia, GSS also became associated with the common neuropathological denominator of numerous, widespread, multicentric amyloid plaques which demonstrate selective immunostaining by anti-PrP antibodies. However, with the passage of time, GSS neuropathology has been associated with a more diverse clinical spectrum, including dementia, spastic paraparesis and even extrapyramidal features which may respond to dopaminergic therapy. The neuropathological phenotype has also been extended to include the presence of neurofibrillary tangles. Divisions based on these varied features may have no intrinsic biological validity given the considerable clinico-pathological diversity described within families.

Epidemiology and Cause

The variety of genetic mutations associated with illnesses classified as GSS since 1989 further complicates the issue, and underscores the likely non-

Table 7. *PRNP* mutations associated with GSS

Missense

Codon	Amino acid change	Codon 129	Codon	Amino acid change	Codon 129
102	proline to leucine	M[1]	198	phenylalanine to serine	V
212	glutamine to proline	?	145	tyrosine to stop	M
217	glutamine to arginine	V	105	proline to leucine	V
117	alanine to valine	V			

[1] This mutation has also been reported in *cis* with valine at codon 129.

Inserts

8-octapeptide repeat (192 bp) insert starting at codon 84
8-octapeptide repeat (192 bp) insert starting at codon 76
8-octapeptide repeat (192 bp) insert, specific site not stated but located within codon 51–91 region

specificity of this diagnosis. This genetic heterogeneity may partly explain the described considerable phenotypic heterogeneity (table 7). Recent reports continue to highlight intra-pedigree phenotypic variation, with different members within one family having illnesses varying from typical CJD to classic GSS [33–41]. The normal polymorphism at codon 129 may also influence phenotype [42, 43]. The likely causative role of these mutations is strongly supported by the spontaneous development of spongiform degeneration in the brains of transgenic mice harbouring such changes [44].

In most instances, a trans-generational neurologic syndrome is apparent within an affected kindred, usually following autosomal dominant transmission patterns, although uncommonly, GSS occurs sporadically [45, 46]. The first described and most common mutation occurs at codon 102 (P102L) which has recently been confirmed in members from the original family described by Gerstmann. This mutation is usually associated with the more typical 'ataxic' GSS clinico-pathological profile but exceptions have been reported, attributed to the modifying effects of non-pathogenic polymorphisms. The normal *PRNP* gene codes for a nonapeptide sequence which with only very minor nucleotide modifications is repeated as four octapeptides between codons 51–91. Inserts of varying multiples of these octapeptide repeats are commonly found in other familial prion diseases, especially CJD; the GSS phenotype appears linked to various 8-octapeptide repeat expansions of this gene region (table 7).

Transmission studies, utilising both non-human primates and rodents, have generally shown low success rates, with the most common causal mutation (P102L) transmissible in approximately 40% of cases; the other less frequent mutations have either not been assessed or did not transmit [47, 48].

Clinical Features

In the more typical GSS cases, symptoms begin in the fifth or sixth decade, but the onset may be as young as 25 years. The illness duration ranges from 3 months to 13 years, with the mean around 5–6 years. Although early complaints are often vague or non-specific there is inexorable, usually slow, progression, so that eventually patients come to manifest differing combinations of cerebellar, pyramidal, behavioural, and cognitive difficulties.

Features of pancerebellar dysfunction typically include: gait unsteadiness, limb ataxia with dysmetria, dysdiadochokinesis and intention tremor, titubation, nystagmus, and dysarthria, often accompanied by dysphagia. Although signs of corticospinal tract degeneration, such as limb weakness, spasticity, hyperreflexia and positive Babinski's responses are present in many patients, in some families (especially with the P105L mutation) spastic paraparesis may dominate the clinical syndrome in the absence of cerebellar dysfunction. The unusual combination of absent lower limb tendon reflexes with extensor plantar responses is relatively frequent in later stages of the illness. Extrapyramidal motor disturbances are common and take the form of adventitious movements (such as myoclonus and athetosis) and rigidity.

Memory impairment is usually the first indication of cognitive decline and with the passage of time a more pervasive dementia invariably manifests. Along with the intellectual deterioration, changes in demeanour are common, ranging through aggression, irritability and emotional lability to apathy and withdrawal. Less common features are deafness, cranial nerve palsies and seizures. Death in a bedridden, akinetic mute, totally dependent state is a frequent culmination.

An American (Indiana) kindred, associated with the F198S mutation, was remarkable for both the development of dopaminergic-responsive parkinsonian features (with minimal or absent tremor) and the presence of neocortical neurofibrillary tangles (NFTs) [34]. A Swedish family with later-onset cognitive decline and cerebellar dysfunction (associated with the Q217R mutation) also manifested neocortical NFTs in addition to the widespread PrP-positive plaques [34]. As already mentioned, the P105L mutation is recognised to cause a familial spastic paraparesis-dementia variant, without clinical evidence of cerebellar dysfunction or myoclonus [38]. An uncommon 'amber' stop mutation, causing termination of translation at codon 145 (Y145stop), was found in a patient with isolated progressive dementia spanning 20 years, who at autopsy had the combination of Alzheimer changes and numerous neocortical and cerebellar PrP-positive amyloid plaques [39]. Finally, the A117V mutation is associated with another 'telencephalic' variant of GSS wherein progressive dementia is the predominant clinical feature, often coexisting with lesser pyramidal and extrapyramidal findings but with minor cerebellar dysfunction [49].

Diagnosis

Neuro-imaging (both CT and MRI scanning) may be normal [50] or show non-specific atrophy affecting the cerebral hemispheres and/or cerebellum. Most often, the EEG only shows a non-specific excess of slower frequencies (which may appear in bursts) but can be normal even in relatively advanced disease [50]. In addition, the generalised, 1- to 2-Hertz triphasic or periodic slow-wave complexes most often sought to aid the diagnosis of sporadic CJD can be seen, usually in patients with more aggressive disease and shorter total illness durations, often less than 12 months [40].

Fatal Familial Insomnia

FFI is a genetically determined transmissible spongiform encephalopathy with a relatively characteristic clinico-pathological phenotype. Although FFI was first applied as a descriptive diagnosis in 1986 to describe an illness afflicting 5 members of a large Italian family [51], it was not until 1992 that the disorder was shown to be a genetically determined prion disease [52], and within a few years confirmed to be transmissible [53].

Epidemiology and Cause

Since its clarification as a prion disorder, a number of additional pedigrees manifesting FFI have been described [54–57]. However, it remains a very rare disorder. Consistent with its characterisation as a D178N mutation in the PRNP, FFI shows inheritance patterns consistent with autosomal dominance [52, 54]. However, incomplete penetrance is possibly quite common as 11 members of the original large Italian FFI kindred were found to harbour the mutation (3 older than 60 years) and yet were asymptomatic at the time of the report. Within FFI pedigrees, onset is usually in the fifth decade, but ranges from 20 to 63 years, with illness durations averaging around 13–15 months (range 6–42 months) [52, 54–58].

In a retrospective genetico-pathological analysis of families previously given the generic diagnostic label 'selective thalamic degeneration', 3 of 4 demonstrated the same D178N *PRNP* mutation, prompting their more correct classification as FFI [59]. As already mentioned, the same D178N mutation had previously been described as a cause of familial CJD in a number of unrelated pedigrees, stimulating discussion as to explanations for this apparent clinico-genetic dichotomy [60]. The modifying influence of the normal polymorphism at codon 129 on the mutant allele has been proposed as the explanation, with FFI occurring in individuals with the D178N-129M haplotype and familial CJD occurring in individuals with the D178N-129V genotype [61]. However, detailed studies of kindreds containing the D178N and other mutations have shown sufficient clinico-pathological diversity and overlap to suggest that FFI and CJD most likely represent somewhat artificial but frequently identifiable clinical phenotypes, and cast doubt on the claim that ultimate illness is governed entirely by associated codon 129 allelic polymorphisms [54, 55, 57, 62–64]. Recently, clinico-pathological phenocopies of FFI unassociated with an underlying PRNP mutation (known as sporadic fatal insomnia) have been described [65, 66].

Clinical

The core clinical features of FFI consist of profound disruption of the normal sleep-wake cycle (with complete disorganisation of the electroencephalographic patterns of sleep), sympathetic overactivity, diverse endocrine abnormalities (particularly attenuation of the normal circadian oscillations) and markedly impaired attention.

The sleep disturbance may initially be relatively minor but usually progresses over weeks to months until normal sleep may not be possible, supplanted by stupor usually accompanied by frequent, vivid dreams which may be acted upon while still somnolent [51]. Prompt arousal with light stimuli remains characteristic but not invariable [57], and as cognition fails, patients may not be able to recall their intrusive dreams. A variety of probable (auditory, visual and tactile) hallucinations may occur in addition to the parasomnias, and further contribute to the bizarre nocturnal behaviours and oneiric automatisms which can be observed [54, 56].

Dysautonomia constitutes the other major distinguishing clinical feature of FFI, and may be noticeably episodic. Although often reflecting sympathetic overactivity, its broader manifestations include: impotence, sphincteric dysfunction, salivation, rhinorrhoea, lacrimation, hyperthermia, hyperhidrosis, tachycardia and hypertension [51, 64]. Autonomic dysfunction tends to occur relatively early in the clinical course and may be the presenting symptom.

As the illness evolves, a range of motor abnormalities usually evolve in variable combinations. Cerebellar and pyramidal dysfunction culminates in prominent limb, gait and bulbar difficulties, accompanied by hyperreflexia, upgoing plantar responses, intention tremor and dysmetria. Spontaneous and reflex myoclonus are commonly present. Disorders of ocular movement and generalised hypertonia may also be seen. Respiration is frequently altered and may display tachypnea, or an irregular noisy pattern with intermittent apnea and hypoventilation [51].

Cognitive impairment usually develops later in the evolution of the illness, but may remain relatively mild and restricted to mnestic difficulties on formal neuropsychological testing [51, 64]. Invariably, patients eventually become confused and disoriented, progressing to stupor and coma with death from intercurrent pneumonia a common outcome. Seizures are not commonly seen during the course of typical FFI.

Diagnosis

Routine biochemical and hematological parameters are typically normal as are CSF findings, although oligoclonal banding of uncertain significance and relevance has been reported in a single patient [56].

Systematic monitoring of FFI patients typically discloses a range of hormonal irregularities, comprising alterations in basal blood levels, changes in the circadian pattern of secretion, or both [51, 56]. Serum cortisol is continuously increased, with or without preservation of circadian fluctuations. Circulating levels of both melatonin and thyrotropin (TSH) are reduced, with greatly attenuated or absent variations over a 24-hour period. Despite this, thyroid hormone levels are reported as normal and TSH levels increase in response to challenges with thyrotropin-releasing hormone. In addition, normal circadian oscillations are lost for growth hormone, prolactin, and follicle-stimulating hormone (FSH), with impaired responses of FSH and luteinising hormone (LH) to LH-releasing hormone. Basal gonadal hormones may be increased (progesterone) or decreased (testosterone and estradiol), but can be partly stimulated artificially with human chorionic gonadotropin.

Polysomnographic recordings confirm markedly reduced total sleep time and gross electroencephalographic disorganisation of sleep, including virtual absence of typical rapid eye movement (REM) periods and deeper non-REM phases characterised by K-complexes, spindles and slow waves [51, 56]. Instead, non-wakefulness may be replaced by something approximating poorly developed REM phases which coincide with periods of dreaming. Even pharmacologic agents such as benzodiazepines and barbiturates may be unable to induce sleep-like EEG activity [51], but promising results were reported in a single patient given γ-hydroxybutyrate [56].

The routine EEG is often normal in the early stages, but usually shows the progressive development of widespread, non-specific slower (theta and delta) frequencies as the disorder advances [51, 54, 56, 57, 64]. However, generalised less responsive alpha activity has been reported during the course of a patient's illness [51]. Periodic complexes or triphasic waves are usually not seen in typical FFI associated with the D178N-129M haplotype [51, 54, 56, 57, 64].

Kuru

Kuru constitutes a horizontally transmitted prion disease of declining incidence, geographically confined to Papua New Guinea. It was first identified by Western medicine in the mid-1950s, as the more remote parts of New Guinea came under external control through the provision of medical and other services [67, 68]. 'Kuru' in the Fore language means 'to shiver' (or 'to be afraid'), and along with cerebellar ataxia, constitutes the salient clinical hallmarks of the disorder.

Epidemiology and Cause

After it was identified, kuru was recognised to be endemic amongst the Fore linguistic and cultural group resident in the Eastern Highlands of New Guinea [67, 69]. It had been extant since at least 1941 but was also seen in the neighbouring Keiagana, Kanite, Kimi, Usurufa and Auiyana tribes with whom the Fore often inter-married. Hence, kuru occurred in a circumscribed, remote, mountainous region populated by approximately 17, 000 indigenes dispersed over more than 160 villages. Since the outlawing of ritualistic endo-cannibalism in the late 1950s, kuru has shown a steady decline and is now almost eradicated [70, 71]. However, at the time of its original description, kuru was estimated to have a general prevalence of approximately 1% within the geographically circumscribed area centred around the Okapa Patrol Post, but an annual incidence of up to 10% was seen in some tribes and accounted for over 50% of all deaths in certain communities.

Over three-quarters of the originally described kuru victims were adult women and children of either sex but symptoms never commenced prior to 4 years of age; adult males only rarely developed kuru. Over the first decade of detailed investigation, the prevalence of kuru declined significantly, with a progressive increase in the youngest age of onset in children, such that whereas the illness was initially observed in children as young as 4 years in 1957, by 1967, symptom onset was never seen under 14 years of age. Eventually, the confluence of scientific and epidemiological data culminated in the theory that kuru had been transmitted and sustained at endemic proportions, by cannibalistic rituals observed as part of the funeral rites mourning deceased relatives [70, 72]. Women and children at these ceremonies ate the less desirable (including the highly infectious central nervous system) tissues while adult males partook of the (relatively non-infectious) organs, such as skeletal muscle. In addition to ingestion of infectious tissues, conjunctival, nasal and skin contamination were other likely modes of transmission. Many observations purport incubation periods of greater than 2, and up to 4 decades [69, 73]. Although the ultimate cause of kuru remains uncertain, a cannibalistic serial passage of a sporadic case of indigenous CJD remains the most plausible hypothesis.

Despite its experimental transmissibility, there is no epidemiological evidence to support vertical transmission, with pregnant symptomatic women (at various stages of kuru) typically delivering healthy infants who, to this date, have never gone on to manifest the illness.

Clinical Features

The onset is typically insidious without antecedent illness, and the overall clinical picture is extremely uniform, predominantly manifesting as an inexorably progressive pancerebellar dysfunction [67–69]. Fever and seizures are not features and the general physical examination is normal until late in the disease when effects of under-nutrition may become evident. Total illness duration is usually 6–9 months, but ranges from 4 to 24 months, with homozygosity (particularly for methionine) at the polymorphic codon 129 of *PRNP* apparently linked to younger age at onset and shorter duration of illness [74].

Subtle ataxia affecting gait and disequilibrium usually herald the first or ambulant phase and may be initially appreciated by family or friends rather than the patient. Once the patients cannot ambulate unassisted due to their progressive cerebellar dysfunction, they are considered to have entered the second or sedentary phase of the illness. Nystagmus is rarely evident but a convergent strabismus frequently occurs, usually late in the clinical course. Within a matter of weeks or a few months, all voluntary motor activity is so impaired that the patients cannot even sit unsupported and are then described as having entered the third or terminal phase during which they are bedridden, totally dependent for feeding and all personal care, and typically incontinent of urine and faeces. Decubitus ulcers are a common complication, and by this stage the patients' bulbar functions have usually deteriorated to the point where they are anarthric and aglutic, with inanition an invariable feature. Death quickly ensues as a consequence of static bronchopneumonia, infected pressure sores or starvation.

Irregular and coarse 'tremors' soon accompany the limb and gait ataxia and may give rise to body actions resembling shivering. The tremors are usually only observed during voluntary motor activity and also often take the form of an intention tremor affecting purposeful appendicular movements and/or titubation of the head and trunk. Additional involuntary movements having the appearance of chorea or athetosis may be seen although some authors ascribe this adventitious motor activity to misunderstood features of cerebellar ataxia [75]. Significant features of pyramidal or extrapyramidal dysfunction (akinesia and rigidity) are usually not prominent. Muscle weakness and wasting are not seen until the secondary effects of malnutrition supervene, and somatic sensory functions appear maintained throughout the course of the illness.

Cognition tends to be relatively spared, at least until late in the course of the illness when its accurate assessment is very difficult due to the incapacitating impairments of motor function, including speech. The prevailing mood tends to be one of euphoria through the early phases of the illness and emotional lability and a pseudobulbar effect with inappropriate excesses of laughter or crying may be seen. Ultimately, the patient's demeanour becomes one of apathetic withdrawal.

Diagnosis

As for the other prion diseases, routine biochemical and haematological investigations are normal, as is the CSF [67], with the clinical diagnosis appropriately restricted to indigines of the Eastern Highlands of New Guinea manifesting a typical illness. Neuropathological examination is required for unequivocal confirmation.

References

1   Katscher F: It's Jakob's disease, not Creutzfeldt's. Nature 1998;393:11.
2   Meyer A, Leigh D, Bagg CE: A rare presenile dementia associated with cortical blindness (Heidenhain's syndrome). J Neurol Neurosurg Psychiatry 1954;17:129–133.
3   Gibbs CJ, Gajdusek DC, Asher DM, Alpers HP, Beck E, Daniel PM, Matthews WB: Creutzfeldt-Jakob disease (spongiform encephalopathy): Transmission to the chimpanzee. Science 1968;161:388–389.
4   National CJD: Surveillance Unit, London School of Hygiene & Tropical Medicine. Creutzfeldt-Jakob disease surveillance in the United Kingdom – Eighth Annual Report. 2000.
5   Will RG, Alperovitch A, Poser S, Pocchiari M, Hofman A, Mitrova E, De Silva R, D'Alessandro M, Delasnerie-Laupretre N, Zerr I, van Duijn C: Descriptive epidemiology of Creutzfeldt-Jakob disease in six European countries, 1993–1995. Ann Neurol 1998;43:763–767.
6   Wientjens DPWM, Davanipour Z, Hofman A, Kondo K, Matthews WB, Will RG, van Duijn CM: Risk factors for Creutzfeldt-Jakob disease: A reanalysis of case-control studies. Neurology 1996;46:1287–1291.
7   Collins S, Law MG, Fletcher A, Boyd A, Kaldor J, Masters CL: Surgical treatment and risk of sporadic Creutzfeldt-Jakob disease: A case-control study. Lancet 1999;353:693–697.
8   Harries-Jones R, Knight R, Will RG, Cousens S, Smith PG, Matthews WB: Creutzfeldt-Jakob disease in England and Wales, 1980–1984: A case-control study of potential risk factors. J Neurol Neurosurg Psychiatry 1988;51:1113–1119.
9   Brownell B, Oppenheimer DR: An ataxic form of subacute presenile polioencephalopathy (Creutzfeldt-Jakob disease). J Neurol Neurosurg Psychiatry 1965;28:350–361.

10   Parchi P, Giese A, Capellari S, Brown P, Schulz-Schaeffer W, Windl O, Zerr I, Budka H, Kopp N, Piccardo P, Poser S, Rojiani A, Streichemberger N, Julien J, Vital C, Ghetti B, Gambetti P, Kretzschmar H: Classification of sporadic Creutzfeldt-Jakob disease based on molecular and phenotypic analysis of 300 subjects. Ann Neurol 1999;46:224–233.

11   McNaughton HK, Will RG: Creutzfeldt-Jakob disease presenting acutely as stroke: An analysis of 30 cases. Neurol Infect Epidemiol 1997;2:19–24.

12   Steinhoff BJ, Racker S, Herrendorf G, Poser S, Grosche S, Zerr I, Kretzschmar H, Weber T: Accuracy and reliability of periodic sharp wave complexes in Creutzfeldt-Jakob disease. Arch Neurol 1996;53:162–165.

13   Zerr I, Bodemer M, Gefeller O, Otto M, Poser S, Wiltfang J, Windl O, Kretzschmar HA, Weber T: Detection of 14-3-3 protein in the cerebrospinal fluid supports the diagnosis of Creutzfeldt-Jakob disease. Ann Neurol 1998;43:32–40.

14   Finkenstaedt M, Szudra A, Zerr I, Poser S, Hise JH, Stoebner JM, Weber T: MR imaging of Creutzfeldt-Jakob disease. Radiology 1996;199:793–798.

15   Advisory Committee on Dangerous Pathogens: Transmissible spongiform encephalopathy agents: Safe working and the prevention of infection. London, Stationery Office, 1998.

16   Brown P, Preece MA, Will RG: 'Friendly fire' in medicine: Hormones, homografts, and Creutzfeldt-Jakob disease. Lancet 1992;340:24–27.

17   Healy DL, Evans J: Creutzfeldt-Jakob disease after pituitary gonadotrophins. BMJ 1993;307:517–518.

18   Heckmann JG, Lang CJG, Petruch F, Druschky A, Erb C, Brown P, Nuendorfer B: Transmission of Creutzfeldt-Jakob disease via a corneal transplant. J Neurol Neurosurg Psychiatry 1997;63: 388–390.

19   Hogan RN, Cavanagh HD: Transplantation of corneal tissue from donors with diseases of the central nervous system. Cornea 1995;14:547–553.

20   Will RG, Alpers MP, Dormont D, Schonberger LB, Tateishi J: Infectious and sporadic prion diseases; in Prusiner SB (ed): Prion Biology and Diseases. New York, Cold Spring Harbour Laboratory Press, 1999, pp 465–507.

21   Will RG, Ironside JW, Zeidler M, Cousens SN, Estibeiro K, Alperovitch A, Poser S, Pocchiari M, Hofman A, Smith PG: A new variant of Creutzfeldt-Jakob disease in the UK. Lancet 1996;347: 921–925.

22   Knight R: The relationship between new variant Creutzfeldt-Jakob disease and bovine spongiform encephalopathy. Vox Sang 1999;76:203–208.

23   Hill AF, Butterworth RJ, Joiner S, Jackson G, Rossor MN, Thomas DJ, Frosh A, Tolley N, Bell JE, Spencer M, King A, Al-Sarraj S, Ironside JW, Lantos PL, Collinge J: Investigation of variant Creutzfeldt-Jakob disease and other human prion diseases with tonsil biopsy samples. Lancet 1999; 353:183–184.

24   Will RG: New variant Creutzfeldt-Jakob disease. Biomed Pharmacother 1999;53:9–13.

25   Will RG, Stewart G, Zeidler M, Macleod MA, Knight RSG: Psychiatric features of new variant Creutzfeldt-Jakob disease. Psychiatr Bull 1999;23:264–267.

26   Zeidler M, Stewart GE, Barraclough CR, Bateman DE, Bates D, Burn DJ, Colchester AC, Durward W, Fletcher NA, Hawkins SA, Mackenzie JM, Will RG: New variant Creutzfeldt-Jakob disease: Neurological features and diagnostic tests. Lancet 1997;350:903–907.

27   Zeidler M, Sellar RJ, Collie DA, Knight R, Stewart G, Macleod MA, Ironside JW, Cousens S, Colchester AFC, Hadley DM, Will RG: The pulvinar sign on magnetic resonance imaging in variant Creutzfeldt-Jakob disease. Lancet 2000; 355:1412–1418.

28   Will RG, Zeidler M, Stewart GE, Macleod MA, Ironside JW, Cousens SN, Mackenzie J, Estibeiro K, Green AJE, Knight RSG: Diagnosis of new variant Creutzfeldt-Jakob disease. Ann Neurol 2000;47:575–582.

29   Hill AF, Zeidler M, Ironside J, Collinge J: Diagnosis of new variant Creutzfeldt-Jakob disease by tonsil biopsy. Lancet 1997;349:99–100.

30   Gerstmann J: Über ein noch nicht beschriebenes Reflexphänomen bei einer Erkrankung des zerebellaren Systems. Wine Med Wochenschr 1928;78:906–908.

31   Gerstmann J, Sträussler E: Über eine eigenartige hereditär-familiäre Erkrankung des Zentralnervensystems. Zugleich ein Beitrag zur Frage des vorzeitigen lokalen Alterns. Z Neurol 1936;154:736–762.

32   Masters CL, Gajdusek DC, Gibbs CJ: Creutzfeldt-Jakob disease virus isolations from the Gerstmann-Sträussler syndrome with an analysis of the forms of amyloid plaque deposition in the virus-induced spongiform change. Brain 1981;104:559–588.

33   Hsiao K, Baker H, Crow TJ, Poulter M, Owen F, Terwilliger JD, Westaway D, Ott J, Prusiner SB: Linkage of a prion protein missense variant to Gerstmann-Sträussler syndrome. Nature 1989;338: 342–345.

34   Hsiao K, Dlouhy SR, Farlow MR, Cass C, Da Costa M, Conneally PM, Hodes ME, Ghetti B, Prusiner SB: Mutant prion proteins in Gerstmann-Sträussler-Scheinker disease with neurofibrillary tangles. Nat Genet 1992;1:68–71.

35   Doh-ura K, Tateishi J, Sasaki H, Kitamoto T, Sakaki Y: Pro→Leu change at position 102 of prion protein is the most common but not the sole mutation related to Gerstmann-Sträussler syndrome. Biochem Biophys Res Commun 1989;163:974–979.

36   Goldgaber D, Goldfarb L, Brown P, Asher DM, Brown WE, Lin S, Teener JW, Feinstone SM, Rubenstein R, Kascsak RJ, Boellaard JW, Gajdusek DC: Mutations in familial Creutzfeldt-Jakob disease and Gerstmann-Sträussler-Scheinker's disease. Exp Pathol 1989;106:204–206.

37   Kitamoto T, Ohta M, Doh-ura K, Hitoshi S, Terao Y, Tateishi J: Novel missense variants of prion protein in Creutzfeldt-Jakob disease or Gerstmann-Sträussler syndrome. Biochem Biophys Res Commun 1993;191:709–714.

38   Kitamoto T, Amano N, Terao Y, Nakazato Y, Isshiki T, Mizutani T, Tateishi J: A new inherited prion disease (PrP-P105L mutation) showing spastic paraparesis. Ann Neurol 1993;34:808–813.

39   Kitamoto T, Iizuka R, Tateishi J: An amber mutation of prion protein in Gerstmann-Sträussler syndrome with mutant PrP plaques. Biochem Biophys Res Commun 1993;192:525–531.

40   Goldfarb L, Brown P, Vrbovská A, Baron H, McCombie WR, Cathala F, Gibbs CJ, Gajdusek DC: An insert mutation in the chromosome 20 amyloid precursor gene in a Gerstmann-Sträussler-Scheinker family. J Neurol Sci 1992;111:189–194.

41   Hainfellner J, Brantner-Inthaler S, Cervenakova L, Brown P, Kitamoto T, Tateishi J, Diringer H, Liberski P, Regele H, Feucht M, Mayr N, Wessely P, Summer K, Seitelberger F, Budka H: The original Gerstmann-Sträussler-Scheinker family of Austria: Divergent clinicopathological phenotypes but constant PrP genotype. Brain Pathol 1995;5:201–211.

42   Ghetti B, Dlouhy S, Giaccone G, Bugiani O, Frangione B, Farlow MR, Tagliavini F: Gerstmann-Sträussler-Scheinker disease and the Indiana kindred. Brain Pathol 1995;5:61–75.

43   Young K, Clark H, Piccardo P, Dlouhy SR, Ghetti B: Gerstmann-Sträussler-Scheinker disease with the PRNP P102L mutation and valine at codon 129. Mol Brain Res 1997;44:147–150.

44   Hsiao K, Scott M, Foster D, Groth DF, DeArmond SJ, Prusiner SB: Spontaneous neurodegeneration in transgenic mice with mutant prion protein. Science 1990;250:1587–1590.

45   Yamada M, Itoh Y, Fujigasaki H, Naruse S, Kaneko DK, Kitamoto T, Tateishi J, Fomo E, Hayakawa M, Tanaka J, Matsushita, Miyatake T: A missense mutation at codon 105 with codon 129 polymorphism of the prion protein gene in a new variant of Gerstmann-Sträussler-Scheinker disease. Neurology 1993;43:2723–2724.

46   Liberski P, Barcikowska M, Cerven?kov? L, Bratosiewicz J, Marczewska M, Brown P, Gajdusek DC: A case of sporadic Creutzfeldt-Jakob disease with a Gerstmann-Sträussler-Scheinker phenotype but no alterations in the PRNP gene. Acta Neuropathol 1998;96:425–430.

47   Brown P, Gibbs CJ, Rodgers-Johnson P, Asher D, Sulima MP, Bacote A, Goldfarb LG, Gajdusek DC: Human spongiform encephalopathy: The National Institutes of Health series of 300 cases of experimentally transmitted disease. Ann Neurol 1994;35:513–529.

48   Tateishi J, Kitamoto T, Hoque MZ, Furukawa H: Experimental transmission of Creutzfeldt-Jakob disease and related diseases to rodents. Neurology 1996;46:532–537.

49   Hsiao K, Cass C, Schellenberg GD, Bird T, Devine-Gage E, Wisniewski H, Prusiner SB: A prion protein variant in a family with the telencephalic form of Gerstmann-Sträussler-Scheinker syndrome. Neurology 1991;41:681–684.

50   Brown P, Goldfarb L, Brown WT, Goldgaber D, Rubenstin R, Kascsak RJ, Guiroy DC, Piccardo P, Boellaard JW, Gajdusek DC: Clinical and molecular genetic study of a large German kindred with Gerstmann-Sträussler-Scheinker syndrome. Neurology 1991;41:375–379.

51   Lugaresi E, Medori R, Montagna P, Baruzzi A, Cortelli P, Lugaresi A, Tinuper P, Zucconi M, Gambetti P: Fatal familial insomnia and dysautonomia with selective degeneration of thalamic nuclei. N Engl J Med 1986;315:997–1003.

52   Medori R, Tritschler H-J, LeBlanc A, Villare F, Manetto V, Chen H, Xue R, Leal S, Montagna P, Cortelli P, Tinuper P, Avoni P, Mochi M, Baruzzi A, Hauw J, Ott J, Lugaresi E, Autilio-Gambetti L, Gambetti P: Fatal familial insomnia, a prion disease with a mutation at codon 178 of the prion protein gene. N Engl J Med 1992;326:444–449.

53   Tateishi J, Brown P, Kitamoto T, Hoque ZM, Roos R, Wollman R, Cervenáková L, Gajdusek DC: First experimental transmission of fatal familial insomnia. Nature 1995;376:434–435.

54   McLean C, Storey E, Gardner RJM, Tannenberg AEG, Cervenáková L, Brown P: The D178N (cis-129M) fatal familial insomnia mutation associated with diverse clinicopathologic phenotypes in an Australian kindred. Neurology 1997;49:552–558.

55   Medori R, Montagna P, Tritschler HJ, LeBlanc A, Cortelli P, Tinuper P, Lugaresi E, Gambetti P: Fatal familial insomnia: A second kindred with mutation of prion protein gene at codon 178. Neurology 1992;42:669–670.

56   Reder A, Mednick A, Brown P, Spire JP, van Cauter E, Wollmann RL, Cervenáková L, Golfarb LG, Garay A, Ovsiew F, Gajdusek DC, Roos RP: Clinical and genetic studies of fatal familial insomnia. Neurology 1995;45:1068–1075.

57   Silburn P, Cervenáková L , Varghese P, Tannenberg A, Brown P, Boyle R: Fatal familial insomnia: A seventh family. Neurology 1996;47:1326–1328.

58   Gambetti P, Parchi P, Petersen RB, Chen SG, Lugaresi E: Fatal familial insomnia and familial Creutzfeldt-Jakob disease: Clinical, pathological and molecular features. Brain Pathol 1995;5: 43–51.

59   Petersen R, Tabaton M, Berg L, Schrank B, Torack RM, Leal S, Julien J, Vital C, Deleplanque B, Pendelbury WW, Drachman D, Smith TW, Martin JJ, Oda M, Montagna P, Ott J, Autilio-Gambetti L, Lugaresi E Gambetti P: Analysis of the prion protein gene in thalamic dementia. Neurology 1992;42:1859–1863.

60   Goldfarb L, Haltia M, Brown P, Nieto A, Kovanen J, McCombie WR, Trapp S, Gajdusek DC: New mutation in scrapie amyloid precursor gene (at codon 178) in Finnish Creutzfeldt-Jakob kindred. Lancet 1991;337:425.

61   Goldfarb L, Petersen R, Tabaton M, Brown P, LeBlanc AC, Montagna P, Cortelli P, Julien J, Vital C, Pendelbury WW, Haltia M, Wills RR, Hauw JJ, McKeever, Monari L, Schrank B, Swergold GD, Autilio-Gambetti L, Gajdusek DC, Lugaresi E, Gambetti P: Fatal familial insomnia and familial Creutzfeldt-Jakob disease: Disease phenotype determined by a DNA polymorphism. Science 1992;258:806–808.

62   Bosque P, Vnencak-Jones C, Johnson MD, Whitlock JA, McClean MJ: A PrP gene codon 178 base substitution and a 24-bp interstitial deletion in familial Creutzfeldt-Jakob disease. Neurology 1992; 42:1864–1870.

63   Chapman J, Brown P, Goldfarb LG, Arlazoroff A, Gajdusek DC, Korczyn AD: Clinical heterogeneity and unusual presentations of Creutzfeldt-Jakob disease in Jewish patients with the PRNP codon 200 mutation. J Neurol Neurosurg Psychiatry 1993;56:1109–1112.

64   Manetto V, Medori R, Cortelli P, Montagna P, Tinuper P, Baruzzi A, Rancurel G, Hauw J, Vanderhaeghen J, Mailleux P, Bugiani O, Tagliavini F, Boras C, Rizzuto N, Lugaresi E, Gambetti P: Fatal familial insomnia: Clinical and pathologic study of five new cases. Neurology 1992;42: 312–319.

65   Mastrianni JA, Nixon R, Layzer R, Telling GC, Han D, DeArmond SJ, Prusiner SB: Prion protein conformation in a patient with sporadic fatal insomnia. N Engl J Med 1999;340:1630–1638.

66   Parchi P, Capellari S, Chin S, Schwarz HB, Schecter NP, Butts, Hudkins P, Burns DK, Powers JM, Gambetti P: A subtype of sporadic prion disease mimicking fatal familial insomnia. Neurology 1999;52:1757–1763.

67   Gajdusek D, Zigas V: Degenerative disease of the central nervous system in New Guinea: The endemic occurrence of 'kuru' in the native population. N Engl J Med 1957;257:974–978.

68   Zigas V, Gajdusek D: Kuru: Clinical study of a new syndrome resembling paralysis agitans in natives of the Eastern Highlands of Australian New Guinea. Med J Austr 1957;2:745–754.

69   Liberski P, Gajdusek D: Kuru: Forty years later, a historical note. Brain Pathol 1997;7:550–560.
70   Gajdusek D: Unconventional viruses and the origin and disappearance of kuru. Science 1977;197: 943–960.
71   Klitzman R, Alpers M, Gajdusek DC: The natural incubation period of kuru and the episodes of transmission in three clusters of patients. Neuroepidemiology 1984;3:3–20.
72   Gajdusek D: Kuru: An appraisal of five years of investigation. Eugen Q 1962;9:69–74.
73   Prusiner S, Gajdusek D, Alpers MP: Kuru with incubation periods exceeding two decades. Ann Neurol 1982;12:1–9.
74   Cervenáková L, Goldfarb L, Garruto R, Lee H-S, Gajdusek DC, Brown P: Phenotype-genotype studies in kuru: Implications for new variant Creutzfeldt-Jakob Disease. Proc Natl Acad Sci USA 1998;95:13239–13241.
75   Hornabrook RW: Kuru: Some misconceptions and their explanation. Papua New Guinea Med J 1966;9:11–15.

Dr. Steven Collins, National CJD Registry, Department of Pathology,
The University of Melbourne, Parkville, Victoria 3010 (Australia)
Tel. +61 3 8344 5867, Fax +61 3 8344 4004, E-Mail stevenjc@unimelb.edu.au

Rabenau HF, Cinatl J, Doerr HW (eds): Prions. A Challenge for Science,
Medicine and Public Health System. Contrib Microb. Basel, Karger, 2001, vol 7, pp 93–104

....

Epidemiology and Risk Factors of Transmissible Spongiform Encephalopathies in Man

Inga Zerr, Sigrid Poser

Department of Neurology, University of Göttingen, Germany

A general surge of interest in the incidence of Creutzfeldt-Jakob disease (CJD) is primarily associated with the discovery of the transmissibility of kuru and also with the occurrence of bovine spongiform encephalopathy (BSE). Due to the fear that this cow disease might be transmitted to man through the consumption of beef products, epidemiological studies have been initiated throughout Europe. Prior to these investigations, almost exclusively mortality statistics were available worldwide (table 1). According to these statistics, the total annual number of individuals who died of CJD was between 0.1 and 1 per million population (=mortality). A classification into sporadic, genetic and iatrogenic cases was not possible. So far, a systematic registration of the incidence over longer periods has only been performed by the United Kingdom, where retrospective studies were already started in the seventies [1]. After the occurrence of BSE in 1986, the investigations were continued using modified methods of data acquisition. Former investigations in different parts of the world, which primarily included mortality statistics based on death certificates, revealed an incidence between 0.31 and 0.42 [2]. In subsequent prospective studies, it could be shown that the increase in registered cases – the incidence was meanwhile reported to be 0.74 – was due to the different methodological approach. The improvement of diagnostics and, probably, awareness of CJD within the group of patients aged above 70 years is one factor influencing incidence [3, 4].

Molecular biological procedures have meanwhile enabled the detection of different mutations in the prion protein gene (*PRNP*) and thus a distinction from sporadic cases. In addition, genetic examinations allow the identification of patients who, formerly, would not have been diagnosed as CJD cases owing

Table 1. Annual incidence rates of CJD

Country	Survey years	Incidence (cases per million)
Argentinia	1980–1996	*34 cases*
Australia	1979–1992	0.75
Chile	1955–1972	0.10
	1973–1977	0.31
	1978–1983	0.69
Germany	1979–1990	0.31
	1993–1997	0.85[1]
France	1968–1977	0.34
	1978–1982	0.58
	1992–1995	0.96[1]
United Kingdom	1964–1973	0.09
	1970–1979	0.31
	1980–1984	0.47
	1985–1989	0.46
	1990–1994	0.70[1]
	1995–1996	0.74[1]
India	1971–1990	*30 cases* (0.002)
Iceland	1960–1990	0.27
Israel	1963–1987	0.91
	1989–1997	0.90
Italy	1958–1971	0.05
	1972–1986	0.09
	1993–1995	0.56[1]
Japan	1975–1977	0.45
New Zealand	1980–1989	0.88
Netherlands	1993–1995	0.81[1]
Austria	1994–1995	1.27[1]
Sweden	1985–1996	1.20
Switzerland	1988–1997	1.14
Slovakia	1993–1995	0.62
Czechoslovakia	1972–1986	0.66
Hungary	1960–1986	0.39

Table 1 (continued)

Country	Survey years	Incidence (cases per million)
United States	1973–1977	0.26
	1979–1990	0.90
Byelorussia	1981–1989	*21 cases*

Table was derived from Brown et al. [2], modified by inclusion of published data.

[1] Prospective studies.

Cases of CJD have also been reported from Egypt, Belgium, Brazil, China, Finland, Greece, Indonesia, Iran, Yugoslavia, Canada, Colombia, Northern Ireland, New Guinea, Norway, Mexico, Oman, Peru, Poland, Portugal, Romania, Senegal, Spain, South Africa, Taiwan, Thailand, Tunisia, Uruguay, Venezuela, and West Bengal.

to the atypical clinical symptomatology, as has been shown for various mutations [5]. The occurrence of iatrogenic CJD cases, which have been observed since 1985 and which are caused by contaminated dura mater grafts and growth hormone preparations, has also contributed to an increase in the total incidence. In the course of the BIOMED study in the years 1993–1995 [6], 12% of all cases in France and 6% of the cases in the United Kingdom were of iatrogenic origin, mostly after treatment with growth hormones derived from human cadaveric pituitary glands [6].

Sporadic Creutzfeld-Jakob Disease

Sporadic CJD cases occur in all age groups, but no case has ever been reported in a child. The youngest patient seen in the German CJD surveillance study was 23 and the oldest 88 years of age. Analysis of the age-related incidence in the framework of the European BIOMED study shows an identical peak in the age group 70–79 years in the individual countries. A decline in the incidence of sporadic CJD can be seen in patients above 80 years of age (fig. 1).

The lower incidence of CJD in older patients was already reported in earlier studies and can be explained in several different ways: on the one hand, other dementias are frequent in this age group, and it was often not possible in the past to distinguish CJD from Alzheimer's disease or vascular dementia due to the lack of reliable diagnostic tests. On the other hand, it is also

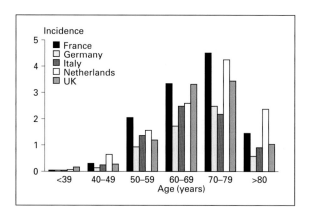

Fig. 1. Age-specific incidence (cases per million per year) of sporadic CJD in Europe 1993–1995.

Table 2. Genotype distribution at codon 129 in definite and probable sporadic CJD in Germany (n = 353)

	Female:male	Ratio
M/M	165:79	2.1:1
M/V	29:25	1.2:1
V/V	32:23	1.4:1

M = Methionine; V = valine.

conceivable that a genetic cause is responsible for the decreased life expectancy. An atypical clinical picture in older patients that does not fulfil the classification criteria and therefore is not recognized as CJD might also be another explanation. An interesting observation resulted from a study conducted in the United States according to which the age-related incidence in the Black population is reported to be lower [7].

In all countries, the disease occurs more frequently in women than in men (1.4:1). Genetic factors, such as codon 129 genotype of the *PRNP* may help understand these findings (table 2).

The median survival time until death in sporadic CJD is 7 months [8], but may also vary to a great extent between 3 weeks and three years. Cases with longer duration were reported occasionally [9]. Patients who are homozygous for methionine or valine on codon 129 of the *PRNP*, tend to have a more rapid disease course than heterozygous patients (fig. 2).

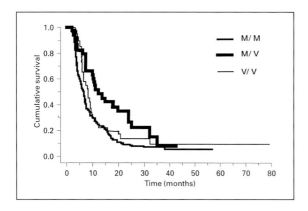

Fig. 2. Kaplan-Meier cumulative survival time stratified by codon 129 genotype. M = Methionine; V = valine.

Table 3. Annual mortality rates per million of sporadic CJD in Europe

	Australia	Austria	France	Germany	Italy	Nether-lands	Slovakia	Spain	Switzer-land	UK
1993	1.01	0.77	0.60	0.44	0.48	0.72	0	0.36	1.44	0.65
1994	0.61	1.28	0.77	0.80	0.58	1.18	0.80	0.38	1.42	0.87
1995	1.04	1.15	1.02	1.00	0.49	0.52	0.40	0.31	1.27	0.60
1996	1.30	1.15	1.19	0.89	0.83	0.83	0.40	0.56	1.41	0.68
1997	1.02	0.77	1.38	1.22	0.76	1.21	0.40	0.51	1.41	1.01
1998	1.07	1.03	1.12	1.27	0.85	1.15	0.40	0.71	1.27	0.87

BIOMED Study

Since 1993, the frequency of the various disease forms of CJD has been registered within the scope of a prospective BIOMED study (table 3). Between 1993 and 1995 data from altogether 592 patients were analyzed throughout Europe. These analyses revealed a comparatively homogenous incidence and mortality of sporadic CJD cases in the individual European countries (table 3; fig. 1).

Risk Factors

Due to the neuropathological similarity between kuru and scrapie [10], a possible link between scrapie and CJD [11] was examined in the past. At least from an epidemiological point of view, the numerous investigations

Table 4. Case-control studies: CJD-associated findings/risk factors

Surgery [50, 51, 54]
Physical injuries [54]
Tonometry within 2 years before onset [51]
Suture age 15 to 3 years before onset [51]
Head-face-neck injury [51]
Other trauma [51]
Family history of dementia (excluding familial cases) [53, 55]
Herpes zoster in adult life [53]
Consumption of raw meat [55, 49] and brain [49]
Exposure to fish, squirrel, rabbit [52]
Exposure to leather products, fertilizer consisting of hoofs and horns [55]
Residence or employment on a farm or market garden [50]

performed revealed no connection: in scrapie-free countries the occurrence of CJD, as compared to those countries in which scrapie is endemic, is equally frequent [11]. Especially in France and Italy a different epidemiological distribution of scrapie and CJD was evidenced [12]. However, these conclusions are subject to the reservation that the worldwide trade with (contaminated) sheep products is not taken into consideration.

Potential risk factors for CJD were examined in several case-control studies (table 4). The methodology of the individual studies is different, in particular with regard to the selection of the control group, which might bias the results [13]. Although significant factors have been found in each study (table 4), the attempt to identify one common environmentally related or endogenous factor failed. In uncontrolled studies some risk factors were proposed to be related to CJD. Owing to the potential transmission of the disease to medical staff, the following case reports deserve special attention: one internist who, 30 years ago, had worked in a department of pathology for a period of 12 months developed CJD, although no direct contact with CJD was documented [14]. One pathologist [15], one neurosurgeon [16], and two assistants who worked in a histology department [17, 18] developed CJD. Furthermore, one orthopedist is reported to have performed experiments with sheep dura mater 30 years prior to disease onset [19]. In spite of these reports, no increased CJD disease risk was revealed by controlled studies in medical professions [20].

The consumption of brain tissue from various free-ranging animals was occasionally associated with the occurrence of sporadic CJD [21, 22]. The observation of an increased occurrence of sporadic CJD in dairy farmers led to the assumption that these cases could be connected with BSE cases in their herds [23]. A comparison of the occupationally related incidence of CJD in

Table 5. Iatrogenic cases of CJD

Mode of infection	Patients n	Agent entry into brain	Mean incubation period (range)	Country of occurrence
Corneal transplant	3	optic nerve	16, 18, 320 months	Germany, Japan, USA
Stereotactic EEG	2	intracerebral	18 months (16, 20)	Switzerland
Neurosurgery (without dura)	4	intracerebral	19 months (12–28)	France, UK
Dura mater graft	106	cerebral surface	6 years (1.5–16)	worldwide, most in Japan
Growth hormone	125	haematogenous	12 years (5–30)	most in France, UK, and USA
Gonadotrophin	5	haematogenous	13 years (12–16)	Australia

Table modified according to Brown [30] and Cambridge Healthtech Institute's Conference on Transmissible Spongiform Encephalopathies, Washington, D.C., October 27–28, 1999.

dairy farmers showed an incidence beyond expectation of about 4 cases per million population per year in other European countries and did not show any difference as compared to the United Kingdom [4]. The occupationally related incidence is higher than the general incidence of CJD, but it is in conformity with the expected number of diseased individuals in the age group 60–70 years. Moreover, clinical symptoms, neuropathological findings and experimental transmission to mice in diseased dairy farmers in the United Kingdom did not differ from sporadic cases [24].

Iatrogenic Cases

Transmission Pathways

A man-to-man transmission of the infectious agent has only been evidenced through direct iatrogenic exposure to infectious tissue (dura mater, cornea, and growth hormones derived from human cadaveric pituitaries, in rare cases by insufficiently sterilized surgical instruments or EEG electrodes; table 5). The first iatrogenic transmission ever reported was recognized only because of the short incubation period in a patient with a corneal graft 18 months before the first symptoms. The donor – as was found out later – had died of CJD. Another case was reported by a Japanese working group, but no data are on hand about the cornea donor [25]. The third known case of an iatrogenic transmission occurred in Germany [26]. Two patients developed

CJD 16 and 20 months following examination with stereotactic EEG electrodes, which had been used in a patient with CJD and were insufficiently sterilized [27]. The infectivity of these electrodes has been demonstrated by transmission in chimpanzees [28]. Owing to the use of insufficiently sterilized neurosurgical instruments, the infectious agent was transmitted in 4 further cases [29, 30].

The transmission of CJD via dura mater grafts by lyophilized dura from a single manufacturer is well documented. Prior to 1987, no additional inactivation procedures with NaOH had been performed [31, 32]. More than 100 cases were reported worldwide [Brown, pers. commun.], most of them from Japan [31]. In the study period, 1 of 3,000 recipients of dura mater grafts developed CJD in that country. Only few cases were seen after the use of noncommercial products [33]. Disease transmission also occurred after non-neurosurgical procedures following orthopedic surgery [31], cholesteatoma [34], after embolization of a nasopharyngeal angiofibroma [35], and after that of an intercostal artery [36].

Up to 1985, patients with primary hypopituitarism were treated with human growth hormones derived from human cadaveric glands. At that time, the first cases of CJD occurred in young patients who had received intramuscular injections of pituitary-gland-derived hormones [37, 38]. The mean incubation time in these patients is 12 years (age range: 5–30 years) [30]. More than 100 cases have meanwhile become known worldwide, 62 of these in France [39], 32 in the United Kingdom, and 28 in the United States [Brown, pers. commun.]. According to estimations, about half of the preparations produced in France between January 1984 and March 1985 might have been contaminated [39].

The incubation time of the disease depends on the portal of entry of the infectious agent which is shorter after intracerebral infection (several months, dura mater), but may range from years to decades after peripheral inoculation (12 years, growth hormones). Genetic factors (genotype at codon 129 of the *PRNP*) influence the incubation time. While so far primarily growth hormone-infected individuals with a methionine or valine homozygosity (shorter incubation time) have developed CJD, the number of persons with a methionine/valine heterozygosity is now steadily increasing (longer incubation time) [40, 41].

New Variant Creutzfeldt-Jakob Disease

The number of cases of sporadic CJD in the United Kingdom has remained stable over the observation periods or has increased slightly due to methodological reasons (fig. 3). In the course of a co-operation between epidemiological working groups of the BIOMED study, altogether 78 patients (75 in the United Kingdom, 2 in France and 1 in Ireland) with a new variant of

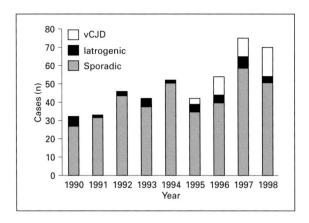

Fig. 3. CJD in the UK.

Table 6. Mortality from vCJD in the UK

Year	Deaths
1995	3
1996	10
1997	10
1998	16
1999 (–Oct.)	8

Creutzfeldt-Jakob disease (vCJD) have been identified so far (table 6) [Will, pers. commun., November 1999; 42, 43]. These cases differ clinically and histopathologically from the classic sporadic form of CJD in the young age at disease onset (median 28 years, range 14–52), early behaviorial abnormalities, depression and anxiety [44], frequent dysesthesia, late dementia and longer disease duration (14 months). Neuropathological lesions of the brain are characterized by pronounced plaque-like prion protein depositions in the entire brain, showing a dense eosinophil center surrounded by spongiform changes (so-called florid plaques), which are not seen in sporadic CJD. The epidemiological link of the novel CJD phenotype in the UK to BSE was soon substantiated by experimental transmission data [24, 45].

The possible incubation time in these patients can only be assessed. Presumably, it is at least 5–10 years, since the exposure to the BSE agent was highest in the 1980s and the first disease cases were observed in 1994. Some recent data point toward an increase in the numbers of new cases at the end of 1998 [46]. However, considering the currently available data, an exact prognosis regarding the further development of the vCJD cases is speculative. Depending on the postulated incubation time and transmission probability [47], the numbers range from some hundred to 80,000 diseased individuals. The risk factors remain unknown. Regarding life history, eating habits, or occupational exposure, no particulars have been reported for the vCJD cases observed in the United Kingdom [48]. Both cases registered in France have never been to the United Kingdom [43]. In contrast, the female patient with vCJD in Ireland lived in the UK between 1989 and 1992 and is considered to have been exposed to BSE there.

References

1   Will RG, Matthews WB, Smith PG, Hudson C: A retrospective study of Creutzfeldt-Jakob disease in England and Wales 1970–1979. II. Epidemiology. J Neurol Neurosurg Psychiatry 1986;49:749–755.
2   Brown P, Cathala F, Raubertas RF, Gajdusek DC, Castaigne P: The epidemiology of Creutzfeldt-Jakob disease: Conclusion of a 15-year investigation in France and review of the world literature. Neurology 1987;37:895–904.
3   Will RG: Incidence of Creutzfeldt-Jakob disease in the European Community; in Gibbs CJ Jr (ed): Bovine Spongiform Encephalopathy: The BSE Dilemma. New York, Springer, 1996, pp 364–374.
4   Cousens SN, Zeidler M, Esmonde TF, De Silva R, Wilesmith JW, Smith PG, Will RG: Sporadic Creutzfeldt-Jakob disease in the United Kingdom: Analysis of epidemiological surveillance data for 1970–96. BMJ 1997;315:389–395.
5   Windl O, Giese A, Schulz-Schaeffer W, Zerr I, Skworc K, Arendt S, Oberdieck C, Bodemer M, Poser S, Kretzschmar HA: Molecular genetics of human prion diseases in Germany. Hum Genet 1999;105:244–252.
6   Will RG, Alpérovitch A, Poser S, Pocchiari M, Hofman A, Mitrova E, de Silva R, D'Alessandro M, Delasnerie-Lauprêtre N, Zerr I, van Duijn C: Descriptive epidemiology of Creutzfeldt-Jakob disease in six European countries, 1993–1995. EU Collaborative Study Group for CJD. Ann Neurol 1998;43:763–767.
7   Holman RC, Khan AS, Kent J, Strine TW, Schonberger LB: Epidemiology of Creutzfeldt-Jakob disease in the United States, 1979–1990: Analysis of national mortality data. Neuroepidemiology 1995;14:174–181.
8   Poser S, Zerr I, Schulz-Schaeffer WJ, Kretzschmar HA, Felgenhauer K: Die Creutzfeldt-Jakob-Krankheit. Eine Sphinx der heutigen Neurobiologie. Dtsch Med Wochenschr 1997;122:1099–1105.
9   Brown P, Rodgers Johnson P, Cathala F, Gibbs CJJ, Gajdusek DC: Creutzfeldt-Jakob disease of long duration: Clinicopathological characteristics, transmissibility, and differential diagnosis. Ann Neurol 1984;16:295–304.
10 Hadlow WJ: Scrapie and kuru. Lancet 1959;ii:289–290.
11 Masters CL, Harris JO, Gajdusek DC, Gibbs CJJ, Bernoulli C, Asher DM: Creutzfeldt-Jakob disease: Patterns of worldwide occurrence and the significance of familial and sporadic clustering. Ann Neurol 1979;5:177–188.
12 Chatelain J, Cathala F, Brown P, Raharison S, Court L, Gajdusek DC: Epidemiologic comparisons between Creutzfeldt-Jakob disease and scrapie in France during the 12-year period 1968–1979. J Neurol Sci 1981;51:329–337.
13 Zerr I, Brandel J-P, Masullo C, Wientjens D, de Silva R, Zeidler M, Granieri E, Sampaolo S, van Duijn C, Delasnerie-Lauprêtre N, Will RG, Poser S: European surveillance on Creutzfeldt-Jakob disease: A case-control study for medical risk factors. J Clin Epidemiol 1999, in press.

14 Berger JR, David NJ: Creutzfeldt-Jakob disease in a physician: A review of the disorder in health care workers. Neurology 1993;43:205–206.

15 Gorman DG, Benson DF, Vogel DG, Vinters HV: Creutzfeldt-Jakob disease in a pathologist. Neurology 1992;42:463.

16 Schoene WC, Masters CL, Gibbs CJJ, Gajdusek DC, Tyler HR, Moore FD, Dammin GJ: Transmissible spongiform encephalopathy (Creutzfeldt-Jakob disease). Atypical clinical and pathological findings. Arch Neurol 1981;38:473–477.

17 Miller DC: Creutzfeldt-Jakob disease in histopathology technicians. N Engl J Med 1988;318:853–854.

18 Sitwell L, Lach B, Atack E, Atack D, Izukawa D: Creutzfeldt-Jakob disease in histopathology technicians. N Engl J Med 1988;318:854.

19 Weber T, Tumani H, Holdorff B, Collinge J, Palmer M, Kretzschmar HA, Felgenhauer K: Transmission of Creutzfeldt-Jakob disease by handling of dura mater (letter). Lancet 1993;341:123–124.

20 van Duijn CM, Delasnerie-Lauprêtre N, Masullo C, Zerr I, Masullo C, de Silva R, Wientjens DPWM, Brandel J-P, Weber T, Bonavita V, Zeidler M, Alpérovitch A, Poser S, Granieri E, Hofman A, Will RG: Case-control study of risk factors of Creutzfeldt-Jakob disease in Europe during 1993–95. European Union (EU) Collaborative Study Group of Creutzfeldt-Jakob disease (CJD). Lancet 1998;351:1081–1085.

21 Berger JR, Waisman E, Weisman B: Creutzfeldt-Jakob disease and eating squirrel brains. Lancet 1997;350:642.

22 Kamin M, Patten BM: Creutzfeldt-Jakob disease: Possible transmission to humans by consumption of wild animal brains. Am J Med 1984;76:142–145.

23 Almond JW, Brown P, Gore SM, Hofman A, Wientjens PW, Ridley RM, Baker HF, Roberts GW, Tyler KL: Creutzfeldt-Jakob disease and bovine spongiform encephalopathy: Any connection? BMJ 1995;311:1415–1421.

24 Bruce ME, Will RG, Ironside JW, McConnell I, Drummond D, Suttie A, McCardle L, Chree A, Hope J, Birkett C, Cousens S, Fraser H, Bostock CJ: Transmissions to mice indicate that 'new variant' CJD is caused by the BSE agent. Nature 1997;389:498–501.

25 Uchiyama K, Ishida C, Yago S, Kurumaya H, Kitamoto T: An autopsy case of Creutzfeldt-Jakob disease associated with corneal transplantation. Dementia 1994;8:466–473.

26 Heckmann JG, Lang CJ, Petruch F, Druschky A, Erb C, Brown P, Neundorfer B: Transmission of Creutzfeldt-Jakob disease via a corneal transplant. J Neurol Neurosurg Psychiatry 1997;63:388–390.

27 Bernoulli C, Siegfried J, Baumgartner G, Regli F, Rabinowicz T, Gajdusek DC, Gibbs CJ: Danger of accidental person-to-person transmission of Creutzfeldt-Jakob disease by surgery (letter). Lancet 1977;i:478–479.

28 Gibbs C Jr, Asher DM, Kobrine A, Amyx HL, Sulima MP, Gajdusek DC: Transmission of Creutzfeldt-Jakob disease to a chimpanzee by electrodes contaminated during neurosurgery. J Neurol Neurosurg Psychiatry 1994;57:757–758.

29 Will RG, Matthews WB: Evidence for case-to-case transmission of Creutzfeldt-Jakob disease. J Neurol Neurosurg Psychiatry 1982;45:235–238.

30 Brown P: Environmental causes of human spongiform encephalopathy; in Baker HF, Ridley RM (eds): Prion Diseases. Totowa, Humana Press, 1996, pp 139–154.

31 Centers for Disease Control and Prevention. Creutzfeldt-Jakob disease associated with cadaveric dura mater grafts – Japan, January 1979–May 1996. JAMA 1998;279:11–12.

32 Diringer H, Braig HR: Infectivity of unconventional viruses in dura mater (letter). Lancet 1989;i:439–440.

33 Pocchiari M, Masullo C, Salvatore M, Genuardi M, Galgani S: Creutzfeldt-Jakob disease after non-commercial dura mater graft (letter). Lancet 1992;340:614–615.

34 Thadani V, Penar PL, Partington J, Kalb R, Janssen R, Schonberger LB, Rabkin CS, Prichard JW: Creutzfeldt-Jakob disease probably acquired from a cadaveric dura mater graft. Case report. J Neurosurg 1988;69:766–769.

35 Antoine JC, Michel D, Bertholon P, Mosnier JF, Laplanche JL, Beaudry P, Hauw JJ, Veyret C: Creutzfeldt-Jakob disease after extracranial dura mater embolization for a nasopharyngeal angiofibroma. Neurology 1997;48:1451–1453.

36   Defevre L, Destee A, Caron J, Ruchoux MM, Wurtz A, Remy J: Creutzfeldt-Jakob disease after an embolization of intercostal arteries with cadaveric dura mater suggesting a systemic transmission of the prion agent. Neurology 1997;48:1470–1471.

37   Koch TK, Berg BO, De AS, Gravina RF: Creutzfeldt-Jakob disease in a young adult with idiopathic hypopituitarism: Possible relation to the administration of cadaveric human growth hormone. N Engl J Med 1985;313:731–733.

38   Gibbs CJJ, Asher DM, Brown PW, Fradkin JE, Gajdusek DC: Creutzfeldt-Jakob disease infectivity of growth hormone derived from human pituitary glands (letter). N Engl J Med 1993;328:358–359.

39   Huillard d'Aignaux J, Alperovitch A, Maccario J: A statistical model to identify the contaminated lots implicated in iatrogenic transmission of Creutzfeldt-Jakob disease among French human growth hormone recipients. Am J Epidemiol 1998;147:597–604.

40   Deslys JP, Jaegly A, d'Aignaux JH, Mouthon F, de Villemeur TB, Dormont D: Genotype at codon 129 and susceptibility to Creutzfeldt-Jakob disease (letter). Lancet 1998;351:1251.

41   Huillard d'Aignaux J, Costagliola D, Maccario J, Billette de Villemeur T, Brandel JP, Deslys JP, Hauw JJ, Chaussain JL, Agid Y, Dormond D, Alpérovitch A: Incubation period of Creutzfeldt-Jakob disease in human growth hormone recipients in France. Neurology 1999;53:1197–1201.

42   Will RG, Ironside JW, Zeidler M, Cousens SN, Estibeiro K, Alperovitch A, Poser S, Pocchiari M, Hofman A, Smith PG: A new variant of Creutzfeldt-Jakob disease in the UK. Lancet 1996;347: 921–925.

43   Chazot G, Broussolle E, Lapras C, Blattler T, Aguzzi A, Kopp N: New variant of Creutzfeldt-Jakob disease in a 26-year-old French man. Lancet 1996;347:1181.

44   Zeidler M, Johnstone EC, Bamber RW, Dickens CM, Fisher CJ, Francis AF, Goldbeck R, Higgo R, Johnson-Sabine EC, Lodge GJ, McGarry P, Mitchell S, Tarlo L, Turner M, Ryley P, Will RG: New variant Creutzfeldt-Jakob disease: Psychiatric features. Lancet 1997;350:908–910.

45   Lasmezas CI, Deslys JP, Demalmay R, Adjou KT, Lamoury F, Dormont D, Robain O, Ironside J, Hauw JJ: BSE transmission to macaques. Nature 1996;381:743–744.

46   Will RG, Cousens SN, Farrington CP, Smith PG, Knight RS, Ironside JW: Deaths from variant Creutzfeldt-Jakob disease (letter). Lancet 1999;353:979.

47   Cousens SN, Vynnycky E, Zeidler M, Will RG, Smith PG: Predicting the CJD epidemic in humans. Nature 1997;385:197–198.

48   Will RG, Knight RSG, Zeidler M, G. S, Ironside JW, Cousens SN, Smith PG: Reporting of suspect new variant Creutzfeldt-Jakob disease. Lancet 1997;349:847.

49   Davanipour Z, Alter M, Sobel E, Asher DM, Gajdusek DC: A case-control study of Creutzfeldt-Jakob disease: Dietary risk factors. Am J Epidemiol 1985;122:443–451.

50   Collins S, Law MG, Fletcher A, Boyd A, Kaldor J, Masters CL: Surgical treatment and risk of sporadic Creutzfeldt-Jakob disease: A case-control study. Lancet 1999;353:693–697.

51   Davanipour Z, Alter M, Sobel E, Asher D, Gajdusek DC: Creutzfeldt-Jakob disease: Possible medical risk factors. Neurology 1985;35:1483–1486.

52   Davanipour Z, Alter M, Sobel E, Asher DM, Gajdusek DC: Transmissible virus dementia: Evaluation of a zoonotic hypothesis. Neuroepidemiology 1986;5:194–206.

53   Harries-Jones R, Knight R, Will RG, Cousens S, Smith PG, Matthews WB: Creutzfeldt-Jakob disease in England and Wales, 1980–1984: A case-control study of potential risk factors. J Neurol Neurosurg Psychiatry 1988;51:1113–1119.

54   Kondo K, Kuroiwa Y: A case control study of Creutzfeldt-Jakob disease: Association with physical injuries. Ann Neurol 1982;11:377–381.

55   van Duijn CM, Delasnerie-Laupêtre N, Masullo C, Zerr I, Masullo C, de Silva R, Wientjens DPWM, Brandel J-P, Weber T, Bonavita V, Zeidler M, Alpérovitch A, Poser S, Granieri E, Hofman A, Will RG: Case-control study of risk factors of Creutzfeldt-Jakob disease in Europe during 1993–95. European Union (EU) Collaborative Study Group of Creutzfeldt-Jakob disease (CJD). Lancet 1998;351:1081–1085.

Inga Zerr, MD, Neurologische Klinik, Georg-August-Universität Göttingen,
Robert-Koch-Strasse 40, D–37075 Göttingen (Germany)
Tel. +49 551 39–6636, Fax +49 551 39–7020, E-Mail 106004,1022@compuserve.com

Rabenau HF, Cinatl J, Doerr HW (eds): Prions. A Challenge for Science,
Medicine and Public Health System. Contrib Microb. Basel, Karger, 2001, vol 7, pp 105–144

....

Bovine Spongiform Encephalopathy and Its Relationship to the New Variant Form of Creutzfeldt-Jakob Disease

An Account of Bovine Spongiform Encephalopathy, Its Cause, the Clinical Signs and Epidemiology Including the Transmissibility of Prion Diseases with Special Reference to the Relationship between Bovine Spongiform Encephalopathy and the Variant Form of Creutzfeldt-Jakob Disease

R. Bradley

Contact address: Veterinary Laboratories Agency, New Haw, Addlestone, UK

Bovine spongiform encephalopathy (BSE or 'mad cow disease') is a new disease. It was first confirmed in the United Kingdom (UK) in November 1986 following microscopic examination of the brains from two cows in southern England [1]. The lesions were consistent with a diagnosis of a spongiform encephalopathy, though at this time it was not known if the disease was transmissible. Nevertheless, the features of these two cases, and several more presented during the course of 1987, showed a remarkable resemblance to the lesions found in scrapie of sheep.

Scrapie, a transmissible spongiform encephalopathy (TSE) of sheep, goats and moufflon was well known in Great Britain and had existed there and in France, Germany and in countries bordering the Danube valley since the early 18th century [2].

In 1985, only six TSEs were known, three in animals (scrapie, transmissible mink encephalopathy of farmed mink and chronic wasting disease of some species of Cervidae) and three in man (Creutztfeldt-Jakob disease, Gerstmann-Sträussler-Scheinker disease and kuru). With the possible exception of scrapie, all are rare diseases. Some are geographically localized: transmissible mink encephalopathy to North America and Northern Europe [3], chronic wasting disease to North America [4] and kuru to the Fore-speaking people in the eastern highlands of Papua New Guinea [5].

In the summer of 1986, also in Great Britain, a scrapie-like disease was identified in an adult captive nyala (*Tragelaphus angasi*) [6]. Subsequently, a total (to 1 October 1999) of 121 cases of scrapie-like spongiform encephalopathy have been reported, mostly in the United Kingdom (UK), in a total of eight species of captive wild Bovidae, five species of captive wild Felidae and in domestic cats. Domestic cats contribute 88 cases to the total [7]. Most, or all, of these incidents can be attributed to probable exposure to the BSE agent from cattle. To appreciate how this may have occurred, see 'Epidemiology and Transmission' sections below.

Returning to BSE, it soon became clear that, like scrapie and other TSEs, the new disease was a prion disease. There were two reasons for this. First was the discovery, at an early stage, that disease-specific fibrils morphologically resembling scrapie-associated fibrils (SAF) were present in the brains of affected cows [1]. SAF are aggregates of a disease-specific neuronal membrane protein (called prion protein, PrP) that can be isolated from detergent extracts of unfixed brain material from the brains of sheep confirmed to have scrapie and also from the brains of other species affected with TSE.

Secondly, it was shown that the major protein of BSE fibrils is the bovine homologue of PrP as judged by its size, protease resistance, immunoreactivity, lectin-binding and partial N-terminal protein sequence [8]. The diagnosis and investigation in the research context of all TSEs have been enhanced by application of refined methods for PrP detection (e.g. by immunoblotting and immunocytochemistry) that have been developed in recent years. Most notable has been the improvement in methods of extraction and concentration of any PrP present and the generation of carefully designed and improved antibodies or techniques to detect PrP at low concentrations. These may enable distinction to be made between the normal cellular form of PrP (PrP^C) and the disease-specific form PrP^{Sc} [9, 10].

In March 1996, 10 cases of a new variant form of CJD (vCJD) were announced in the UK and subsequently the case reports were published [11]. The independent UK Spongiform Encephalopathy Advisory Committee (SEAC) stated that in the absence of any other plausible explanation, the most likely cause was exposure to BSE before the introduction of the specified bovine offals ban in 1989. The specified offals included those tissues from cattle that were thought most likely to contain the BSE agent even before clinical signs were evident. Neither at the time, nor subsequently, has a definite link been made between BSE in cattle and vCJD by virtue of occupation or the eating habits of affected individuals [12]. Thus the precise origin of infection for humans who develop vCJD is unknown.

Measures have been applied to protect animal and human health from BSE. The former (discussed briefly in the 'Epidemiology' section below) has resulted in a decline in the UK epidemic and the disease is en route for elimination in the early years of this century. New exposures of humans to the BSE agent via food have ceased in the UK. Uncertainties about key epidemiological parameters, such as the extent of effective exposure, patient susceptibility and the length of the incubation period, make predictions of the size and duration of the vCJD epidemic difficult [13]. There are currently (20 April 2000) a total of 52 definite or probable cases of vCJD in the UK and 2 in France and 1 in Ireland [14].

In summary, BSE, several other new or newly recognized spongiform encephalopathies of animals and vCJD share the pathological features of TSE and molecular biological features of prion diseases. Diagnostic methods for BSE and vCJD are well developed, but at present confirmation of the diagnosis can only be made post mortem. The BSE epidemic in the UK is declining towards obscurity as enforced measures take their effect. Prediction of the outcome of the vCJD epidemic is more difficult due to absence of key epidemiological information such as the length of the incubation period. The following sections will deal with the cause, clinical signs and epidemiology of BSE. Finally there follow sections on the transmissibility of prion diseases in general, BSE and vCJD in particular and the relationships between them.

Cause of Bovine Spongiform Encephalopathy

*The Properties of Agents That Cause Transmissible
Spongiform Encephalopathy*

There is still dispute about the nature of the agents that cause TSE [15]. The first point to make is that they are unconventional. The unconventionality of the TSE agents is related to the fact that hosts mount no conventional immune response to infection and that the agents are extraordinarily resistant to physical and chemical inactivation using methods that are usually lethal to conventional bacteria, viruses and fungi. The former property means there is no currently available practical test to detect infected cattle with BSE or humans with vCJD. In regard to the latter property, the infectivity is not indestructible. Methods that reduce infectivity to undetectable or negligible levels have been developed and are practically effective [16, 17]. These include incineration (e.g. for high-risk waste), autoclaving (e.g. for surgical instruments), the use of sodium hypochlorite or sodium hydroxide (e.g. for infected premises or for instruments that cannot withstand autoclaving) and rendering (e.g. for low-risk animal waste). Conditions are specified for each kind of treatment and these must be strictly adhered to.

The Nature of Transmissible Spongiform Encephalopathy Agents

The agents that cause TSE are variously described as prions, virinos or unconventional viruses [15] (see elsewhere in this book for more detailed information). Whatever is their nature, four things are certain.

First, they differ markedly from conventional micro-organisms like viruses, bacteria, fungi and parasites in their unconventionality (see above). However, they share with these organisms two important features: the ability to mutate and strain-specific properties.

The second difference is that infectivity is very closely associated with PrP, though not all disease-specific PrP is infectious.

Thirdly, if the prion hypothesis [18] in its purest form is correct, then the agent is an infectious protein that is host specific and generated post-translationally from the normal cellular form, PrP^C, that is in turn generated by the host's *PrP* gene. Two hypotheses exist to explain the replication of prions [19]. One is template refolding where PrP^C is post-translationally altered by re-folding part of its α-helix, coil structure into β-sheet by an unknown mechanism to produce the partially protease-resistant and insoluble disease-specific form PrP^{Sc}. Differences in conformation and possibly glycosylation may contribute strain-specific properties. The other hypothesis is that a nucleation process is responsible for replication.

Fourthly, some forms of CJD are familial and are inherited in conventional Mendelian fashion. However, the agents isolated from such patients (and generated presumably by the patients' *PrP* gene) are transmissible, just as they are from the sporadic and other forms of CJD and TSE in general. Thus this form of disease is both inherited and transmissible.

The Bovine Spongiform Encephalopathy Agent

Disease results from an interaction of the environment (in this case the agent and its genome if it has one) and the genome of the patient.

Simply put, and in regard to the environmental component, the cause of BSE is exposure to an unconventional scrapie-like agent or prion, known as the 'BSE agent'. The strain of agent appears very stable and has been consistently isolated from around a dozen cattle with the natural disease in the UK, including two cattle from Switzerland [20], and also from a nyala, a greater kudu and three domestic cats, all with TSE. It appears that there is a single major strain of agent in cattle and the biological strain type is different from any of the twenty or so strains of scrapie agent derived from natural cases of scrapie in sheep or from laboratory animals following passage [21]. In this regard, it is unique and has a distinct biological 'fingerprint' that enables field or experimental isolates to be identified with certainty. This fingerprint is determined by inoculating specific in-bred strains of laboratory mice and F1 hybrids and determining the incubation period and lesion profile. The lesion profile is determined by microscopic examination of nine grey matter and three white matter areas of brain for evidence of vacuolation and assessing the severity of this vacuolation. However, this is a lengthy and costly procedure that is not practical for everyday use, so quicker and cheaper molecular methods are being developed, though none is yet validated. The molecular methods are based upon the structure of PrP^{Sc}, notably the fragment size and glycosylation ratios following immunoblotting of this protein.

In regard to the host genetic component in BSE, this has been investigated in a number of countries. For example, although polymorphisms in the bovine *PrP* gene have been reported [22–25], none has been associated with BSE occurrence. At present, therefore, it can be concluded that variation in the sequence of the *PrP* gene in cattle is not a major risk factor for BSE in cattle, rather, the disease is associated only with exposure to the infectious agent probably via the oral route. How this occurs will be discussed in detail in the section on epidemiology.

Clinical Signs of Bovine Spongiform Encephalopathy

A clinical case definition of BSE in the UK has been given by Wilesmith and Wells [26]. The clinical signs have remained constant during the UK epidemic [27] and are closely similar in the different breeds and countries where BSE has occurred [27, 28]. These observations are consistent with the isolation of the same biological strain type of agent from different herds and countries, at least in regard to the UK and Switzerland [20].

The presenting signs may be subtle changes in behaviour such as seeking solitude, a changed order of entry into the milking parlour and hind limb gait ataxia. Occasionally, the presenting sign is recumbency that increases the difficulty of making a clinical diagnosis because recumbency (the so-called 'downer cow') is a common feature of dairy practice and associated with a wide range of causes [29, 30]. Furthermore in the UK, because over 70% of dairy cows calve between July and December there can be false associations with a seasonal occurrence of disease. The initial clinical signs may develop when cows are dry, separated from the lactating herd and out at grass during the summer months, so are less closely and frequently observed. They are stressed less and initial signs may not be sufficiently clear to provoke suspicion of BSE. But signs become increasingly obvious with time and especially when they calve and rejoin the lactating herd when observation is increased. This results in an increase in reporting at this time.

The stress of transportation can also reveal clear clinical signs not observed before the event. There is also evidence from studies in the research environment that when removing or reducing the stress of lactation and standardizing the husbandry and feeding in a stress-free way, clinical signs may subside. However, given time, the signs usually do progress to become obvious, provided the suspect is kept for long enough.

The clinical signs of BSE are similar to those of scrapie and fall into the same categories of altered sensation, mental status, posture and locomotion. As in scrapie, the signs are insidious in onset, are progressive and lead to a fatal outcome. The predominating signs are neurological and include apprehensive behaviour, hyperaesthesia and gait ataxia. The duration of signs ranges from 7 days to 14 months, but is typically 1–2 months. This helps to distinguish most other acute, infectious diseases, including rabies. As BSE is a notifiable disease in all countries where it occurs, the full duration of the clinical phase is foreshortened by the need to compulsorily slaughter the animal once the disease is suspected or a clinical diagnosis has been made.

Sensation

The most frequently recorded signs are apprehension, changes in temperament and behaviour, abnormal ear position, nervousness of entrances, nervous ear and eye movements, teeth grinding and frenzy.

Mental Status

The most frequently recorded signs are hyperaesthesia to touch, sound and light, excessive licking of the muzzle and flanks and head shyness. Simple tests can be applied to determine if individual cows have increased sensitivity to sound (e.g. banging a metal tray) or to light (putting the animal in a darkened room and then switching on the light). Kicking in the milking parlour is also a common sign. This can be very violent if the lower hind limbs are touched, whether in the parlour or not. It is unwise to examine a suspect case of BSE alone and without adequate protection or restraint.

Posture and Locomotion

In regard to posture, low head carriage, laidback ears, arched back and a wide-based stance at rest are common at-rest features. The most frequently recorded signs during movement are hypermetria, gait ataxia and falling (especially on slippery concrete and when turning). Permanent recumbency usually precedes death.

Signs positively correlated with BSE include hypersensitivity to touch or sound, teeth grinding, apprehension, kicking and ataxia. Negatively correlated signs are circling and blindness [31].

General signs include loss of bodily condition and weight, and reduced milk yield. The loss of bodily condition and weight might be attributed, at least in part, to reduced rumination that is very evident in cattle with BSE [32]. Bradycardia has been reported in cattle with BSE [33, 34]. Automated, remote methods have been adapted for measuring the heart rate in cattle suspected to have BSE, and their diagnostic usefulness has been confirmed [35]. Some cattle show disturbances in cardiac rhythm. By administration of atropine, it has been shown that the bradycardia is mediated by increased vagal influence, suggesting that the cardio-inhibitory reflexes in the caudal brainstem are functionally altered in BSE [35]. Healthy cattle deprived of food also exhibit bradycardia.

Less frequently encountered signs include drinking by lapping (like at cat) rather than by sucking. Some cattle may get apparent relief from assumed pruritus following rubbing of the middle of the back but the scratch reflex seen in sheep is not a prominent feature.

Epidemiology of Bovine Spongiform Encephalopathy

During 1987, initial epidemiological studies [36] of 200 BSE-affected herds and cases eliminated several possible causes of BSE. These included: direct and indirect contact with sheep (20% of affected farms had had no sheep on them in living memory), imported cattle and semen, vaccines, hormones, herbicides, pesticides (including organo-phosphorous compounds) and other toxic substances. The only common feature was the use, in concentrate rations prepared for cattle, of the end products resulting from the processing of the unwanted products of slaughter and was mostly derived from abattoirs and butchers. The process is called rendering. Although new chemical methods of rendering are beginning to appear, the traditional method involves cooking (with or without added fat) to remove water, followed by pressing or centrifuga-tion to separate the tallow (fat) from the greaves (protein). Meat and bone meal (MBM) is produced from the greaves fraction by grinding. It soon became evident that, of the end products, MBM was the most likely vehicle of infection. This was because the BSE agent was more likely to partition with the protein than the fat fraction and because the use of MBM was more parochial than that of tallow that tended to be processed in a small number of specialist plants. If tallow had been the source there would have been a more even distribution of BSE occurrence [37]. A case-control study supported this view

[38]. Meantime, experimental transmission of BSE to mice from brain material from affected cattle by the autumn of 1988 had established that BSE was unquestionably a TSE [39].

Thus it was concluded that BSE is an infectious prion disease transmitted mainly, if not entirely, via concentrate feed. The route of exposure from feed is presumed to be oral but conjunctival or nasal routes cannot be ruled out entirely given the proximity of the whole head to the feed source during feeding and especially when automatic feeders are used e.g. in milking parlours. The vehicle of infection is MBM produced by rendering of unwanted waste tissues resulting mainly from slaughter.

Rendering of Animal Waste

Some rendering processes use hyperbaric conditions, others operate at atmospheric pressure. Some are continuous processes, others batch processes. The question arose as to why BSE was a new disease apparently caused by a TSE agent in the MBM suddenly evident in the period 1985–1986 when MBM had been used as a feed ingredient for cattle for several decades previously. Indeed in 1950 and onwards in Great Britain MBM was used as a feed supplement in 'national grain balancer feed' and 'national high protein-concentrate feed for cattle' [40]. Although it is difficult to authenticate by documentation, it is believed that MBM was used in cattle feed at least from the 1930s. So why did BSE suddenly appear for the first time in the mid-1980s?

A hypothesis that explained this was advanced by Wilesmith et al. [37], who conducted a detailed survey of rendering plants in Great Britain in 1988. This survey revealed that two major changes had been occurring in the rendering industry during the 1970s and earlier 1980s. The first was a gradual change from continuous to batch processing that had been operating since 1972 and was still continuing. However, the study did not reveal significant differences in the mean maximum temperatures used in the two processes. Thus this change could not itself account for the sudden change in exposure of cattle in 1981–1982, to a dose of BSE agent sufficient to cause disease by the oral route. A sudden change at this time was required to explain the timing of the origin of the epidemic for two reasons. First, the mean incubation period of BSE was calculated to be 60 months [41]. This meant that the first effective exposures (i.e. those that would have resulted in clinical disease had the recipients lived long enough), resulting in the first cattle with BSE in 1985–1986, would have been in 1981–1982.

Second, at the outset, BSE was geographically widely distributed throughout Great Britain. Because MBM was mostly used on farms close to its point of manufacture and as there were approaching 50 plants operating in

1988, there must have been a change that affected either the majority of plants or, strategically located plants that produced a higher than average output of MBM. The survey revealed that a reduction in the use of hydrocarbon solvents for the recovery of extra tallow from MBM resulted in about a 40% reduction in the MBM produced by this method during the critical period. The hypothesis was supported by analysis of the data on the geographical distribution of BSE. This showed that the incidence in Scotland was relatively low compared with that in England and Wales. This was correlated partly with the fact that the only two rendering plants in Scotland used hydrocarbon solvents for extraction of tallow. Cases of BSE in Scotland were attributed mainly to the movement of cattle from south of the border where they had been exposed.

Two other reasons for the variable geographic distribution of BSE as the epidemic approached its peak were as follows. Some rendering plants produced only tallow: the other product, fat-rich greaves, was sent to larger rendering plants where it was added to locally collected raw material for a second rendering so two heat treatments were used and a dilution factor occurred. This presumably had the effect of reducing the amount, or availability, of infected material in the original raw material. There was a negative correlation between the geographical location of such plants producing MBM from this source and the occurrence of BSE in the same region.

The second reason was the variation in the market share of different feed manufacturers that varied the inclusion and amount of MBM in feed prepared for cattle. This was shown following epidemiological analyses of BSE occurrence in the Channel Islands. There was a much higher incidence of BSE in Guernsey than in Jersey. In both islands dairy cows were fed high levels of concentrate feed because there were no milk quotas there (as there were in the rest of the EU) and the milk from these breeds was particularly rich in fat, thus requiring a high dietary input. The variation in BSE incidence between the two islands was positively correlated with the frequency with which MBM was used in the feed prepared by different manufacturers in Great Britain that supplied the feed for the two islands.

It is important to note that the changes in the rendering processes that were adopted by the industry were commercially motivated and not directed by, or consequent upon, government action. The changes were partly stimulated by safety considerations. A large explosion of hydrocarbon solvents similar to those used in rendering had effectively destroyed a factory and part of a village with loss of life, thus sending a clear message about the risks involved. Also, a fall in the price of tallow and rising costs of energy made the process uneconomic for most producers of tallow. Some of the hydrocarbon solvents used were carcinogenic and toxic, so this was another reason to phase them out, even though they were removed by live steam stripping (for re-use) from MBM before sale.

No plausible alternative hypotheses have been put forward. However, experimental studies to investigate, under laboratory conditions, whether the BSE and scrapie agent can be destroyed by hydrocarbon solvents used in rendering, either alone or in combination with dry or wet heat processes, have revealed little (<1 log) loss of infectivity [42]. It was proposed that the heat was more likely than the solvents to have caused the small amount of inactivation that was measured. Thus it was concluded that the hydrocarbon solvent treatment of MBM had little capacity to reduce infectivity. For sound experimental reasons, a mouse-passaged BSE agent (301V) was used in the study rather than the BSE agent itself. The experimental method was very sensitive, but it could just be that the BSE agent from cattle would have behaved differently. In any case, the small reduction in titre might still have been just enough to prevent an effective dose of BSE agent being delivered to cattle, in feed, via MBM produced by this method.

Rendering experiments have been completed and have shown that two processes used in the EU were unable to completely inactivate the BSE agent [43]. These processes were banned from use for processing ruminant protein in 1994. Similar studies using the scrapie agent revealed that only one of the tested processes was effective in inactivating the scrapie agent [44]. This hyperbaric process (133 °C, 3 bar for 20 min or processes providing the same degree of assurance) is now the only one permitted for rendering ruminant waste destined for use as animal feed in the EU [45]. Later in this document, the term 'defective rendering system' is used. In this document, it is used to indicate a process that did not effectively inactivate scrapie and/or BSE agents and implies no deficit in producing final products devoid of conventional agent infectivity, and no deficit of nutritional quality or of inadequate operation of the plant.

Origin of the Epidemic

For BSE to suddenly develop and escalate as it did, at least three factors were required. These included an infective agent that could replicate in cattle tissue, a large population of susceptible cattle and a means of recycling the agent. The recycling component was achieved by a combination of rendering ruminant waste from infected cattle, incorporation of infected MBM into feed and then feeding it back to cattle. Those cattle that were kept long enough to accommodate the full incubation period expressed clinical disease. The source of the agent that initiated BSE in cattle is explained below. Evaluation of the susceptibility of cattle was achieved by analysing the *PrP* gene from healthy control and BSE-affected cattle, and by conducting pedigree analyses.

PrP *Genetics and Influence of Breed on the Occurrence of*
Bovine Spongifrom Encephalopathy

The bovine *PrP* gene has five or six copies of a short, G-C-rich element
within the protein-coding exon [22]. The frequency of occurrence of these
octapeptide repeat polymorphisms in healthy cattle and cattle with BSE in
Scotland revealed no differences [25]. The structure of the *PrP* gene of Belgian
cattle [24] and US cattle [23] seems no different from that of British cattle.
These studies collectively show that the 6:6 repeat is the most frequent, the
6:5 is moderately frequent and the 5:5 repeat is rare. The *PrP* gene of cattle
seems not to exert a detectable influence on the occurrence or incubation
period of BSE. Hau and Curnow [46] studied the incidence of BSE in related
animals and developed a statistical model. They concluded that there was still
no evidence, molecular or statistical, for genetic variation in susceptibility.

The incidence of BSE within dairy breeds is constant. However, the numbers of cases within each dairy breed varies directly with the breed size [47].
BSE predominates in the Holstein Friesian breed because it is numerically by
far the most common dairy breed.

An analysis for beef breeds cannot be done because there are no available
denominators for the numerical size of the breeds. Dairy cattle are predominantly affected by BSE because of the system of feeding. Calves from dairy
breeds are removed from the dam within 24 h or so of birth, fed artificial milk
and are subsequently fed hay and a protein-rich concentrate ration often
containing MBM. Beef calves on the other hand are suckled and are less
frequently fed concentrate rations containing MBM.

Transmission studies of BSE using two different dairy breeds [48] showed
100% susceptibility and closely grouped incubation periods. Furthermore there
has been a consistency in the neuropathology of BSE in all cattle throughout
the epidemic, supporting the view that only one major agent strain is responsible. Collectively, these studies show there is no variation in susceptibility of
cattle as a result of breed or *PrP* genetics.

Descriptive Epidemiology

Data on suspect cases of BSE have been collected throughout the epidemic
by using a questionnaire, amended from time to time in the light of experience.
This has enabled a detailed analysis of the epidemic to be made and an
epidemiological model to be developed. The model permits future predictions
to be made following various interventions (e.g. culling strategies) and the
investigation of problems like the occurrence of cases born after the introduction of the ruminant feed ban (see below).

Once BSE was made a notifiable disease in the UK in 1988, data were
collected on the suspects reported, those reported and then movement-

restricted, those restricted then compulsorily killed, and of these, the number confirmed by microscopic examination of the brain. The last mentioned provide data for confirmed cases of BSE that can be plotted on the epidemic curve by month and year of clinical onset. This is the true scientific representation of the epidemic. Because there is inevitably a delay from the date of onset of clinical signs of any cow actually with BSE and the date when the case is officially confirmed, the epidemic curve is always representing the true picture as it was around 6 months previous to the current date. It is important to note that some reported and restricted animals make a complete clinical recovery and are not cases of BSE. Also, some cases that are compulsorily killed are not confirmed as BSE. Around 40% of the latter have other neuropathological lesions. Around 60% of the BSE-negative cases are presumed to have a metabolic origin as no morphological lesions are seen. The percentage of suspect cases that are histologically negative has on average been around 15%, but the percentage is now increasing as the epidemic declines.

There are seasonal changes in the negative rate. The rate tends to increase in the spring, as at this time of the year there are more cases of metabolic disease after cattle are turned out to grass following winter housing. Signs attributable to metabolic disorders can mimic the signs of BSE, but can often be discriminated by biochemical analysis of blood or milk samples.

Although data from these studies are put into the public domain they are often misunderstood, partly for reasons given above and partly because reports prepared for different purposes refer to cases in the UK, Great Britain or sometimes the British Isles, including the Channel Islands.

The epidemic reached a peak in early 1993 (when the incidence was about 1% of breeding animals) and since then there has been a sustained decline as a result of the introduction of the ruminant feed ban in 1988. The epidemic is expected to decline close to extinction by 2001 [49]. But since the incubation period may be as long as the lifetime of cattle, there could be a few cases occurring for some considerable time into the future. Such cases will be due to historical exposure via feed.

BSE is predominantly a disease of dairy cattle wherever it has occurred. This is because it was associated with the feeding of MBM to calves born into dairy herds and destined for breeding. By October 1997, 60.2% of dairy and 15.9% of beef herds had experienced at least one BSE case [50]. Most cases in beef suckler herds originated as cross-bred calves in dairy herds where they were exposed to infected feed. These female calves were unsuitable as dairy animals and therefore were sold off to beef suckler herds for breeding. Thus there were more purchased cases of BSE in beef suckler than in dairy herds.

The epidemic curve for herds has a similar pattern to the epidemic curve for cases. The incidence of BSE within herds has not altered very much and has always been below 3.6% [50]. However, this is misleading because for the most part in any one year, the significant exposure is to calves in a single birth cohort rather than to the herd as a whole. Wilesmith [51] calculated that on average, if one case of BSE occurred in a birth cohort, this represented an attack rate of 7% in the cohort. It is clear however that some cattle are first exposed to an effective dose as adults. Whereas most cases of BSE occur in cattle aged 4–6 years there is a range, the youngest case being 20 months old and the oldest, 18 years 10 months old.

To explain the low within-herd incidence of BSE that could not be explained by genetic variation in susceptibility, a low average exposure hypothesis has been proposed for BSE [52, 53]. This suggested that the average exposure in affected herds could have been as low as 10 oral LD_{50} of cattle-adapted BSE agent per tonne of concentrate feed. Another feature proposed was that the infectivity was not evenly distributed through concentrate feed but occurred in packets. Thus the incidence of infection within a birth cohort would depend upon the frequency or concentration of packets within the batch. This hypothesis fits well with the other epidemiological observations such as the higher risk of BSE occurrence in larger herds compared with smaller herds. This is because a farmer would have a greater chance of buying an infected packet if he purchased a large amount of feed. On average, larger batches of feed were purchased by farmers with large herds than by those with small herds.

Hypotheses for the Origin of Bovine Spongiform Encephalopathy

Three main hypotheses for the origin of the epidemic have been put forward [37]. The first, that it was initiated by a mutant strain of the scrapie agent from sheep, is not supported by the epidemiological findings. Results from these studies reveal an extended common-source epidemic. Had a mutant strain of scrapie been responsible, it would have to have been geographically widespread which is not plausible. A single mutation, if it had been a source, would have given rise to a point source, propagative epidemic that clearly did not happen. However, some people consider that the first introduction of BSE may have occurred much earlier than is supposed and that the visible epidemic was the result of a second passage of the agent.

The alternative two hypotheses, namely that it was due to infection with a scrapie-like agent from sheep or a cattle-adapted scrapie-like agent, are unresolved.

Scrapie occurs in sheep and goats at an unknown incidence in a wide range of countries especially in the Northern Hemisphere. In some countries like Argentina, Australia and New Zealand, scrapie is believed to be absent.

Cattle and sheep, embryos and semen have been exported to other parts of the world from the UK and European countries over a long period of time. These and some other countries had strict control of imports of live sheep and goats before the BSE era because of the scrapie risk. But, restrictions on the importation of cattle and cattle products because of a TSE risk was not considered until the discovery of BSE. No case of BSE has occurred in Argentina, any other country of South America, Australia, New Zealand or the United States, despite these potential risks. All these countries have very strict import controls that virtually eliminate the risk of importing these diseases.

It is also interesting to note that cattle challenged parenterally with brain material from sheep in the USA with natural scrapie developed a neurological disease that was unlike BSE as reported from the UK [54]. Furthermore there was very little morphological pathology in the brain though PrP was found [54]. This disease – 'experimental scrapie in cows' – has not been produced by challenge by the oral route, neither has it been reported as a natural disease of cattle anywhere in the world.

Factors Influencing the Occurrence of Bovine Spongiform Encephalopathy in the UK

Sheep Source Hypothesis

Since the epidemic started and has been focused on the UK there must have been special features peculiar to that country that enabled the epidemic to be initiated. One factor was the change in rendering ruminant waste as mentioned above, but some other countries used similar methods to those used in the UK. Assuming an origin from sheep that were a reservoir of scrapie infection in Great Britain, it was clear that there was a unique set of parameters operating there. There was a large population of sheep that outnumbered cattle by about 4:1 and this was a much higher ratio than existed in any other country in Europe or North and South America. This population provided a relatively larger proportion of sheep waste for rendering than any other country. Because most plants rendering mammalian waste accepted material from all species, there would have been a high likelihood of sheep origin MBM getting into any feed for cattle in which MBM was incorporated. The next factor was the deficit in rendering that was revealed later, and finally the common use of MBM in the diets of dairy calves destined for breeding.

Assuming an ovine source, the next ready-made step was the recycling of infected cattle tissues through the defective rendering system. This exacerbated the epidemic that then took off in an exponential fashion though it was only revealed later after the incubation period of the already exposed cattle was completely expressed. However, the early epidemiological work establishing ruminant-derived MBM as the vehicle for spreading BSE resulted in the imposition of a statutory ruminant protein feed ban in the UK in 1988. This was at a relatively early stage of the epidemic, and was immediately effective in reducing exposure via feed, though not completely so.

Cattle Source Hypothesis

The alternative potential source of infection from cattle is more difficult to explain though some of the factors influencing the occurrence of BSE, like deficits in rendering processes, are common to this hypothesis too. If cattle were the source there would have to have been a geographically widespread, self-perpetuating infection in cattle that did not, or did not frequently, give rise to clinical disease.

The Case for a 'Sporadic' Bovine Spongiform Encephalopathy Origin. A 'sporadic' form of BSE akin to sporadic CJD operating at the same incidence (1 case per million p.a.) could not have produced sufficient infection in MBM to initiate an epidemic on the scale that was observed [50]. There have been no pathologically confirmed reports of a scrapie-like disease in cattle prior to BSE (1985) anywhere in the world, even after examination of archived adult cattle brain material in several countries, including in the UK, Switzerland [R. Fatzer, pers. commun.], New Zealand [55], Tasmania [56] and Uruguay [57]. In other words, BSE was a new disease.

The Case for a Subclinical Carrier Origin. If one assumed a carrier state of subclinical bovine TSE, this could be a possible cause, but there is no positive evidence for it. One also has to explain why a subclinical carrier state suddenly changed into a clinical disease. There is no evidence at all for a change in the *PrP* gene structure in cattle so the only other potential change could be in the agent itself. There is abundant evidence [58] that scrapie-like agents mutate and strain selection occurs on crossing a species barrier such that a different strain from that in the challenge dose is isolated after successive passages in the new host. Thus, it could be that a previously avirulent strain of BSE agent widely distributed within the cattle population and with a self-perpetuating mode of transmission mutated and was rapidly recirculated via the defective rendering system in the early 1980s. As tissues from cattle infected with the new agent entered the rendering system, the epidemic was increasingly effectively fuelled as explained above.

Conclusion

It is not possible to precisely identify the origin of BSE whether from sheep or cattle. On balance there is more support for a scrapie origin from sheep if only because there was a reservoir of sheep scrapie in the UK long before BSE was observed. The non-establishment of BSE following challenge of US cattle with brain material from US sheep with scrapie could be explained by the fact that the strains of scrapie in the challenge inoculum were not pathogenic for cattle by the oral route. There could well be different strains of scrapie present in British sheep that theoretically could be pathogenic for cattle by the oral route. Experiments to investigate this hypothesis are in progress. It is also clear that rendering processes that did not sufficiently inactivate the agent and the inclusion of MBM in the diets of dairy calves destined to be breeding cattle were critical factors in the initiation of the epidemic. Recycling of cattle infection was also very important in causing an escalation of the epidemic. It should be possible to eliminate BSE (the disease) from cattle if the following measures are in place and enforced. Control of the:

- composition of ruminant feed,
- specification of parameters for rendering that effectively inactivate TSE agents and
- removal and destruction of tissues likely to carry the BSE agent.

Whether these measures will be sufficient to eliminate BSE infection remains to be seen.

Horizontal Maternal and Paternal Transmission

Horizontal Transmission

A case-control study was conducted in 1993 following the recognition of a number of cases of BSE that were born after the introduction of the ruminant feed ban in 1988 [59]. This study revealed that neither horizontal nor maternal transmission could account for the majority of BSE cases born after the ban was introduced. The low within-herd incidence of BSE ($<3.6\%$ throughout the epidemic and declining as the national incidence of BSE declines to 1.7% in 1997 [50]) does not support horizontal transmission being a cause. Furthermore around 30% of herds that have had a case of BSE, have had only one case. Also homebred herds that have had very high incidences within a birth cohort rarely have such high incidences subsequently. All these features enable us to conclude that horizontal transmission is not a factor in BSE.

Maternal Transmission

In regard to maternal transmission, this cannot alone sustain the epidemic in the UK because the necessary contact rate cannot exceed 1:1. That is to say not all affected cows will produce a calf that itself reaches adulthood and also produces a calf. At best, maternal transmission, if it occurred, could only slightly lengthen the epidemic and not prevent eradication of the disease from the UK. A cohort study initiated in 1989 aimed to investigate the occurrence and incidence of BSE in two groups of animals; one group the offspring of BSE-affected cows (the cases) and the other offspring of healthy cattle 6 years old or more (the controls). Each member of a pair came from the same birth cohort and the same farm. These animals were then collected together in pairs and accommodated on three neutral farms where they were all managed and fed identically. Unfortunately, the experiment was confounded by the fact that most, if not all the cattle had potentially been exposed to BSE-infected feed due to the accidental cross-contamination of the diet with MBM. Detailed analyses of the results [60–63] enabled the SEAC to conclude that there is some evidence for direct maternal transmission at a low level. However, variation in genetic susceptibility to BSE following feed-borne exposure might occur. The risk of transmission from dam to calf is likely to be less than 10%. This appears to be confined to animals born after the onset of BSE in the dam and up to 2 years beforehand, with an increasing risk the closer the calf was born to the time of onset of clinical signs in the dam [64].

Paternal Transmission

There is no evidence for the transmission of BSE via the sire either by natural service or via semen. BSE has occurred in around 500 bulls. Semen and reproductive organs from some bulls with clinical BSE have been bioassayed in susceptible mice and found to contain no detectable infectivity [7].

A comprehensive review of the risks of vertical transmission of BSE is included in an Opinion of the Scientific Steering Committee [65].

Cattle with Bovine Spongiform Encephalopathy Born after the 1988 Feed Ban

To 1 October 1999 there have been 175,634 confirmed cases of BSE in Great Britain and 40,118 of these were born after the 1988 feed ban of 18 July 1988, the latest a solitary case born in November 1995. Born-after-the-ban (BAB) cases have occurred in all countries with BSE in native-born cattle with the exception of Belgium [E. Vanopdenbosch, pers. commun.]. In the UK, neither horizontal nor maternal transmission could be responsible for the majority of these cases (see above). The majority of BAB cases of BSE occurring in the approximate 6-month period between mid-July 1988 and the end of 1988 were attributed to the use of ruminant feed manufactured before the ban, but still in the supply chain or on farm. After this date, almost all of this feed would have been used and so another source must have been responsible. At this time, MBM was still legally being used in feed for mono-gastric species in the UK. Thus ruminant feed prepared in mills that also manufactured feeds for monogastric species was potentially at risk from cross-contamination with MBM destined for non-ruminant feed.

A detailed epidemiological study in Great Britain revealed that there was a changing geographical distribution of homebred cases of BSE in the country with time. Whereas the percentage of cases in the Southwest (the region with the largest percentage of cases) was declining in cattle born between 1985/1986 and 1990/1991, the percentage in the Northern and Eastern regions was increasing. These latter regions have relatively much higher pig and poultry populations than other areas, and MBM was used in the diets for these species but not of course for cattle. Following further investigation and the positive detection by an ELISA of ruminant-derived protein in rations prepared for cattle, it became clear that accidental cross-contamination or ruminant rations with MBM was responsible for the continuing, but reduced exposure of cattle to MBM following the 1998 feed ban.

This accidental cross-contamination was believed to occur because common equipment was used in mills preparing feed for ruminants and non-ruminants. The specific sites could be the auger pit and auger at the point of delivery, during the manufacture of the feed, during transportation (i.e. by the use of vehicles contaminated with small amounts of MBM) or on farm.

Another way this could occur was the 'cascade' phenomenon with pelleted feed. It is most important for poultry and pigs that the quality of the pellet is sound. Sometimes, the pellet was not well made and so new feed was manufactured to replace it. The malformed pellets could have been used as ingredients for pig feed, adjusting the recipe as appropriate. Even if the poultry pellets contained MBM this would not be a problem as it was normal practice to include MBM in some feeds for pigs. The same type of 'cascade' could have followed if pellets prepared for feeding to pigs failed the quality checks, but this time the malformed pellets would have more likely been passed into the ruminant feed and may have contaminated it with MBM. In an emergency, farmers might temporarily also have fed non-ruminant rations to cattle when the store of cattle feed had run out before a new delivery, or in inclement weather when access to the ruminant feed in store might have been impossible.

The level of cross-contamination was probably small but the amount involved could have been decisive in permitting a lethal dose of BSE infectivity to be delivered. How small was determined by a quite separate experiment called the 'attack rate' study that was in progress. The objective of this was to determine the attack rate and incubation period of cattle exposed orally to four different doses of untreated brain material from cows with confirmed BSE. The doses were 100 g on three occasions, 100 g on one occasion, 10 g on one occasion and 1 g on one occasion.

The attack rate study is still in progress, but has nevertheless indicated that some cattle in the 1-gram dose group have succumbed to BSE (and of course cattle in the other groups did too but with a higher attack rate and generally a shorter mean incubation period). Although the experiment used brain material from clinical cases, field cases of BSE could only be exposed to brain from sub-clinically affected cattle in the form of MBM, as clinically affected cattle were all incinerated after 8 August 1988. The message was very clear. A very small amount (e.g. 1 g) of fresh brain could contain sufficient infective agent to kill a calf by the oral route 4 years later [G.A.H. Wells, pers. commun.]. This put an entirely different complexion on the cross-contamination issue since relatively trivial amounts of MBM in cattle diets could not be dismissed as harmless. They could carry, if infected, lethal doses of BSE agent. It is now clear that these types of accidental cross-contamination were likely to be the single most important cause of BAB cases of BSE.

Of course, feed for pigs and poultry containing MBM should have carried no infectivity if the ban on specified bovine offals (later specified risk material) had been completely effective. As it turned out, it was not, and thus any deficit there may have been in the rendering process would have allowed infectivity through into the MBM. Once all the weak links in the chain were identified and corrected through modified legislation and improved compliance and enforcement, the feed source of BSE agent disappeared. From 19 March 1996 in the UK, it is prohibited, and becomes a criminal offence, to offer for sale or supply, or to feed to any farm animal species including horses and fish, MBM derived from mammalian species. The effective date of implementation of this Order in the UK, following the completion of a feed recall scheme, was 1 August 1996. No food animal species born after this date in the UK could be exposed to BSE or scrapie agent via MBM. This can now be guaranteed because the law is effectively enforced continuously by a number of checks, and the use of an ELISA and other tests to detect mammalian protein in all food animal feed.

New Variant Form of Creutzfeldt-Jakob Disease and Its Consequences

On 20 March 1996, an announcement was made by the Secretary of State for Health in the UK, of 10 cases of a new variant form of CJD attributed to exposure to BSE. As a result, a ban on exports from the UK of live cattle and cattle products (other than milk) was applied by the European Commission (EC) by the demand of the Member States on 27 March 1996. The UK SEAC advised amongst other things that the feeding of mammalian MBM to all species of farm animals reared for food, including horses and fish, should cease. This ban was in place on 29 March 1996.

Negotiations with EU Member States and the EC eventually resulted in the agreement that exports of deboned meat and meat products from cattle born after 1 August 1996 could recommence from August 1999. This was called the Date Based Export Scheme. The date of 1 August 1996 was chosen because it was the date by which it could be guaranteed that cattle could no longer be exposed to BSE infectivity via feed in the UK. France subsequently challenged the Decision, but the challenge was rejected by the EC on 29 October 1999 following a thorough examination of the French case and the latest data from the UK by the Scientific Steering Committee of the EC.

Geographical Distribution of Bovine Spongiform Encephalopathy

Small numbers of cases of BSE have been confirmed in cattle exported in the pre-clinical phase from the UK to other countries without BSE in their native stock. These countries include (number of confirmed cases), Canada (1), Denmark (1), Falkland Islands (1), Germany (5, + 1 case imported from Switzerland), Italy (2) and Sultanate of Oman (2). Such cases are of little importance so long as they are identified clinically, disease is confirmed by microscopic examination of the brain and the carcase is totally destroyed so that no part can enter any food or feed chain.

Until 1 October 1999, in addition to the BSE cases in the UK noted in the previous section, countries with BSE in native-born cattle, mostly dairy cattle are (cumulative number of confirmed cases): Belgium (9), France (67), Liechtenstein (2), Luxembourg (1), Netherlands (6), Northern Ireland (1,787), other British Isles (1,265), Portugal (321, including 7 imported from the UK), Republic of Ireland (395 including 12 imported from the UK) and Switzerland (308). A very small number of cases of BSE (included in the totals) have been identified in imported animals in countries with BSE in native-born cattle. The origin of their infection is uncertain.

The origin of BSE in all countries with the disease in their native-born stock has been MBM. A Commission Decision of 27 June 1994 prohibited the feeding of mammalian-derived protein to ruminant animals throughout the EU (although some countries had already introduced national legislation to this effect for cattle or all ruminants from 1990). Some of the cases have been attributed to MBM originating in the UK. However, it is possible that MBM prepared in some of these countries from infected cattle tissues could be an origin, especially for the more recently exposed cases. This is because the export of specific bovine offals (SBO) and feed (MBM) containing them or their derived products, to Member States in the EU from the UK was prohibited in 1990.

The epidemic curves (total confirmed cases by year) for five countries are shown in figure 1. It is notable that the shape of the curves (note the different vertical scales) falls into three patterns. The curve for the UK is classical showing a rise then fall in response to the ruminant feed ban. Switzerland and the Republic of Ireland are similar. However in the case of Switzerland, that now reports only a few cases, the curve has risen again. This is of no concern and can be attributed to the recent introduction of additional surveil-

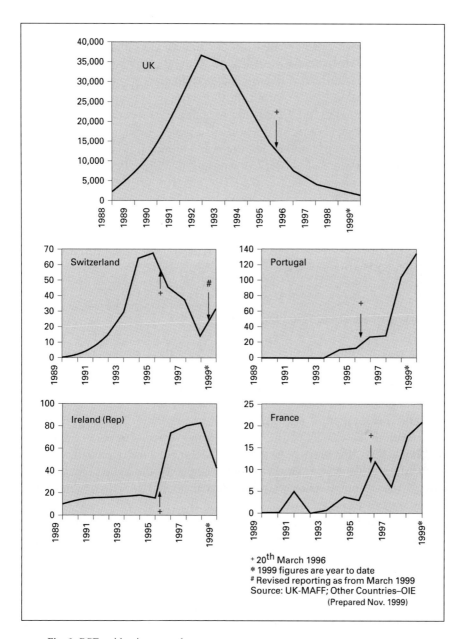

+ 20th March 1996
* 1999 figures are year to date
# Revised reporting as from March 1999
Source: UK-MAFF; Other Countries–OIE
(Prepared Nov. 1999)

Fig. 1. BSE epidemic curves by country.

lance methods. These include the testing of brains from fallen stock, cattle for casualty slaughter and some cattle for human consumption with an approved test for PrP^{Sc} that revealed a small number of additional cases of BSE that were confirmed by microscopic examination of the brain. No other countries have adopted such routine testing officially yet, though several tests have been validated by the EC for use in clinically suspect cattle [66].

The curves for France and Portugal, though indicating much smaller numbers of cases than the UK, are atypical and may not yet have reached a peak. The curves indicate historical exposure to BSE. They do not indicate that cattle are currently being exposed to infected feed, so if European and national laws are enforced, this should be reflected in a downturn in the epidemics in these countries in the future. It is notable that from the date of announcement of 10 cases of vCJD in the UK on 20 March 1996, the curves for the Republic of Ireland, France and Portugal took an upturn, whereas in the UK and Switzerland there was no discernable effect.

Transmissibility of Prion Diseases

In regard to transmission of prion diseases there are five issues to consider. The first is the natural host range and the second the experimental host range. The third issue is the tissue distribution of the agent in the terminal phase of the natural clinical disease. The fourth issue is the distribution and quantitation of infectivity within tissues from the time of exposure until death in the clinical phase. This is usually regarded as part of the study of pathogenesis that seeks to determine also how the infection gets from the point of entry (often the gut in natural disease) to the brain. Finally, there is the issue of biological strain typing and this will be discussed in the last section of this chapter.

Natural Host Range for Bovine Spongiform Encephalopathy
BSE has, or is assumed to have, transmitted from cattle to eight other species of Bovidae, all wild species held captive in zoos and wildlife parks in the UK, to domestic cats [67, 68] and to five captive wild species of Felidae. The cases in Bovidae are presumed to have contracted the disease from the same feed source as cattle. The 1988 ban on feeding ruminant protein to ruminants gave protection to these species and reported cases have declined since, though a small number of BAB cases have occurred in these species too. The cases in captive wild Felidae are presumed to have occurred following consumption of TSE-infected, uncooked carcass material from cattle that contained central nervous tissue [69], probably in the form of bovine heads

(large cats) and vertebral columns (small cats). This practice was prohibited as a result of the SBO ban in 1990. The total number of cases in these species to date has been 33 [MAFF, pers. commun.]. In regard to domestic cats, there has been a total of 86 in the UK with one case each in Norway [70] and Liechstenstein to October 1999. The origin of these cases is more likely to have been from infected MBM or tissues in proprietary pet food but other sources cannot be ruled out. The epidemic curve for cats reached a peak in 1994 (16 cases) but has declined since.

The concurrent appearance of a case of conventional sporadic CJD in a man in Italy and a case of suspected TSE in his pet cat has recently been reported [71]. Assuming the disease in the cat is confirmed to be feline spongiform encephalopathy, it is at present unclear whether this is merely an unusual coincidence of each host acquiring infection from separate sources, or of one host contracting infection from the other or whether each was infected from a common source.

The cause of all the cases in the UK is likely to be infection with the BSE agent that has been formally confirmed in three domestic British cats, the nyala (*Tragelaphus angasi*) and a greater kudu (*Tragelaphus strepsiceros*) following biological strain typing in mice [21, 72]. A scrapie-like spongiform encephalopathy in red-necked ostriches (*Struthio camelus*) in a German zoo has been reported [73, 74]. However, the real nature of this disease is in doubt, particularly as brain material inoculated into experimental rodents susceptible to scrapie did not develop spongiform encephalopathy [H. Diringer, pers. commun.].

Transmissible Spongiform Encephalopathy in Primates Resulting from Exposure to the Bovine Spongiform Encephalopathy Agent

In France, captive primates (a single rhesus monkey (*Macaca mulatta*) [75], macaque monkeys and lemurs [76]) have been reported to have developed a TSE naturally. It is suggested that this has occurred as a result of dietary exposure to TSE infection in feed imported from the UK before controls were in place [76]. Some doubts have been expressed about this because of the atypical and independently unconfirmed clinical and pathological features of the 'natural' disease. Nevertheless, experimental transmission of BSE to some of these primates has been achieved with similar clinical signs and distribution of PrP^{Sc} in the brain. It is important to note that no natural TSE has been reported in any captive primate in the UK, despite at least some of them being fed on feed containing high inclusion rates of MBM derived from UK abattoir waste [77].

These authors were successful in experimentally transmitting BSE and scrapie to each of two marmosets by parenteral inoculation [78]. BSE has also been transmitted experimentally by parenteral routes to cynomolgus macaques [79] and to squirrel and capuchin monkeys [C.J. Gibbs, Jr., pers. commun.]. In regard to the transmission of BSE to man, see below.

Experimental Host Range of the Bovine Spongiform Encephalopathy Agent

So far as is known, BSE has been successfully transmitted to all the species that have been challenged except chickens and hamsters. Successful transmissions by parenteral routes of challenge include: cattle [48], sheep and goats [80], pigs [81], mink [82] and mice [39] (for primates see above). All of these species except pigs are also susceptible by the oral route using high-titre brain material from cattle confirmed to have BSE post mortem.

Tissue Distribution of BSE Infectivity in Cattle with Natural and Experimental Bovine Spongiform Encephalopathy

Over fifty different tissue types from cattle with confirmed BSE, sampled at necropsy of cows in the terminal phase of clinical BSE, have been bioassayed in susceptible mice [7, 83–87, D.M. Taylor, pers. commun.]. Infectivity has been confirmed in central nervous tissues namely the brain, spinal cord (cervical and terminal segments) and the retina. It was not detected in any other tissue including nerves, muscles, milk [84] and male and female reproductive tissues including embryos [86] and semen [7, 83].

Infectivity titres measured in mice have only been reported for brain tissue from cows with confirmed BSE. Despite the indistinguishable clinical signs and extent of the neuropathology in the donor cows, a range of titres has been reported from a maximum $10^{5.2}$ intra-cerebral (i/c) ID_{50} mouse infectious units per gram [88] down to about $10^{3.0}$ mouse i/c ID_{50} per gram, or perhaps even lower. Field cases of disease are now being killed earlier in the clinical phase of disease because of earlier recognition and because of animal welfare considerations. This could account for the lower titres observed in more recent cases.

Sub-passage of BSE in mice (i.e. from cow to mice to mice) has resulted in the selection of a strain called 301V. This can be grown to high titre (10^{10} mouse i/c ID_{50} per gram) in brain and 10^7 in spleen [89].

A more recent investigation of titres in cow brain is incomplete, but nevertheless provides additional information. Due to the species barrier effect there is an obstacle to transmission when crossing a species barrier. This is recognized as a reduced sensitivity of the new species to detect low titres of infectivity. To determine the quantitative difference in sensitivity between mice and cattle for measuring infectivity in cattle a comparative bioassay was set up. This comparative bioassay uses mice and cattle to measure infectivity titres in the pooled brain stem of cows with natural, clinical BSE.

The titration in mice reveals a final titre of $10^{3.3}$ mouse i/c ID_{50}/g of pooled brain stems. The titre in the same brain stem pool when measured in cattle by the i/c route, though incomplete, indicates that it will be in the region of 10^6 cattle i/c ID_{50}/g [G.A.H. Wells, pers. commun.]. Thus it is concluded that the cow has about 1,000 times greater ability (sensitivity) to detect infectivity in cow brain than mice. This is important because most bioassays of tissues for infectivity have used mice that may not detect low levels of infectivity, if present. To overcome this difficulty, certain selected tissues from cattle with experimental BSE have been inoculated into cattle by the i/c route (see next section).

Pathogenesis of Transmissible Spongiform Encephalopathy

Pathogenesis of Experimental Scrapie in Rodents

Much of what we know about the principles of pathogenesis comes from the study of experimental scrapie in mice, e.g. by Kimberlin [90]. Collectively, these studies have indicated that following infection by a peripheral route, there is a lag, zero or eclipse phase following challenge when no infectivity can be detected in any tissue. Replication follows in lymphoreticular system (LRS) tissues like the spleen, lymph nodes, and after oral infection, the organized LRS tissue of the gut. A competent immune system may be required to effect neuroinvasion [91].

Neuroinvasion is a critical phase that in mice is controlled by the *Sinc* (*PrP*) gene. The methods of investigation have involved titration of infectivity in different tissues in experimental scrapie in mice by Kimberlin and Walker [92] in the 1970s and 1980s. Subsequently, the hypothesis advanced, namely that neuroinvasion of the CNS was accomplished via the peripheral nervous system, was supported by Diringer and co-workers [93–95], who detected and measured the amount of PrP^{Sc} in the CNS in experimental hamsters during the incubation period. The results from the investigations of both groups indicate the plausibility and importance of a neural spread of infectivity. A summary is given here.

Following peripheral challenge, infection is first established in the LRS, probably in fixed follicular dendritic cells [96], including in the spleen, lymph nodes and, if the infection is oral or intra-gastric, Peyer's patches of the gut. Lymphocytes may assist fixed follicular dendritic cells to achieve functional maturity and enable replication of the agent. The role of the immune system in prion diseases has been reported [97, 98]. Infection of the CNS is established by several neural routes. In mice one is via the autonomic splanchnic nerve that enters the spinal cord in the mid-thoracic region from whence infection transmits caudally to the distal spinal cord and rostrally to the brain. Another route is via the vagus nerve, which in hamsters may be the only route following oral challenge. Two minor routes in hamsters are via the cervical and thoraco-lumbar cord. When organs are assayed for infectivity during incubation and in the clinical phase, it is clear that infectivity is largely restricted to the LRS and to the CNS, with tissues like muscle being free of detectable infectivity at all times. In the terminal phase of clinical disease the infectivity titre in the CNS exceeds that of the LRS at any time. Very early after exposure, no infectivity in any tissue may be detected. After a lag period of weeks or months (the zero phase), infectivity is detectable in the LRS, and typically detection of infectivity in the CNS is achieved when approximately half the incubation period has elapsed.

Pathogenesis of Natural Scrapie in Sheep

In regard to the pathogenesis of natural scrapie in sheep, seminal studies have been conducted in the USA by William Hadlow and co-workers in the 1970s and early 1980s. They studied goats with natural clinical scrapie [99] and Suffolk [100] and other breeds of sheep [101] with natural clinical scrapie, or incubating the disease. These studies assisted greatly in determining some of the protection measures used in the UK soon after BSE was discovered, notably the selection of tissues (SBO) that should be removed from healthy cattle for destruction. It seems from these studies in Suffolk sheep that there is no major conflict with the pathogenesis as described for rodent models. Deductions about the pathogenesis of scrapie in Suffolk sheep are consistent with those outlined above for experimental scrapie in mice and hamsters. However, in sheep, maternal and horizontal transmissions are important features of the epidemiology of the disease [102]. The placenta harbours infectivity [103–105] and PrP [106] Since maternal transmission of experimental scrapie in the experimental animal species (mice and hamsters) does not occur, the important issues are how the placenta becomes affected in sheep and what tissue element therein harbours this infectivity. There are no published reports of infectivity titres in the placenta of sheep so we do not know how infectious it is in comparison with brain.

This outline of the possible pathogenesis of natural scrapie in sheep is important for the control of scrapie and even more so for the control of BSE should it ever occur in small ruminants. This is because experimental transmission of BSE to sheep has been achieved [80]. In subsequent studies, infectivity was found in both brain and spleen [107]. Since in cattle with natural and experimental BSE no infectivity can be detected in the spleen (see below), it suggests that the pathogenesis of BSE in sheep were it to occur in nature, would be more likely to mimic that of scrapie in sheep than BSE in cattle. As a result, the CNS and certain offals of sheep that contain significant LRS tissue could theoretically contain infectivity in animals incubating the disease, whereas other tissues like muscle, mammary gland and milk do not. Major risk tissues like brain and spleen can be removed in the abattoir and be destroyed so they can enter no food or feed chain. This approach has been taken in some European countries as a precautionary measure as part of a risk reduction policy for the occurrence of BSE in sheep.

Pathogenesis of Experimental Bovine Spongiform Encephalopathy in Cattle following Oral Challenge

Because the within-herd incidence of BSE is very low and because BSE is not a familial disease as is natural scrapie, the pathogenesis of natural BSE could not be determined from the study of natural cases. However, it was important to identify the organs and tissues in cattle incubating BSE that carried infectivity. This was done using an experimental model of BSE in which 30, 4-month-old calves were orally challenged with 100 g of untreated brain tissue from cattle clinically affected with the natural disease and con-firmed to have spongiform encephalopathy post mortem. There were also 10 undosed control cattle. The cattle were sequentially killed at roughly 4-month intervals (three challenged cattle and one control) commencing 2 months after dosing and over forty tissues collected from each for study, the majority of which were bioassayed in susceptible mice. Tissues from each of the three challenged cattle at each 'kill' stage were pooled before inoculation. Clinical signs were not evident in the animals remaining in the challenged group until cattle had reached 35–37 months post-challenge. The shorter period of incubation than in the natural epidemic (mean 60 months) can be attributed to the use of uncooked brain material (not MBM) and the large amount (100 g) and presumed large dose administered.

The results of the above studies have been reported [108–111]. No infectiv-ity was detected in any tissue in cattle killed 2 months after challenge. From 6 to 18 months and from 36 to 40 months infectivity was confirmed in the distal ileum, an area rich in LRS tissue comprised of Peyer's patches. This finding supports the decision to include intestine from duodenum to rectum inclusive in the specified bovine offals bans in various countries, despite not detecting such infectivity in the natural disease. No infectivity was otherwise found in any non-neural tissue other than in a single instance in bone marrow during the clinical phase of disease. Possible explanations for this finding, which include possible cross-contamination have been given [111].

The detection of infectivity in the brain and spinal cord by mouse bioassay, the first occurrence of diagnostic histopathological changes in the brain and the detection of PrP^{Sc} in the brain are essentially coincident at 32 months post-challenge and antedate the onset of clinical signs by 3 months.

Of considerable practical importance was the detection of infectivity in dorsal root ganglia from 32 months post-challenge when the cattle were clinically normal. This finding has resulted in a proposal by the EC to prohibit the use of the vertebral column from cattle (sheep and goats) for the production of mechanically recovered meat throughout the EU from 1 January 2000. Several countries including the UK have already adopted this precaution or gone beyond it.

Because mice are less efficient at detecting BSE than are cattle (see above), some of the tissues collected from the experimental cattle have been inoculated i/c into cattle in order that low levels of infectivity that might be present in tissues 'negative' in mice might be detected. This study is on-going and results are not yet reported.

Relationship of New Variant Creutzfeldt-Jakob Disease to Bovine Spongiform Encephalopathy

On 20 March 1996, as a result of the surveillance activities of the UK CJD Surveillance Unit, the SEAC reported the occurrence of 10 cases of a new variant form of CJD that had affected young people. By March 2000 that number had risen to 52 cases in the UK, 2 cases in France and 1 case in the Republic of Ireland.

Not only is this variant form of CJD (vCJD) new, it is also unusual because of the early age at onset (mean 26 years). Other unusual features are a prolonged clinical phase (just over a year), clinical presentation, progression and neuropathology.

A dozen or so animal species have developed TSE for the first time between 1986 and October 1999. From 3 of these an agent has been isolated with the same biological properties as the BSE agent. Epidemiological investigation showed that it was possible, and indeed likely, that either MBM or uncooked CNS tissues derived from cattle was the source of infection for these species. It was thus very clear that consideration had to be given to the possibility that BSE had arisen in yet another species, *Homo sapiens*, following exposure to the BSE agent. This idea was strengthened because vCJD had followed on soon after the outbreaks of BSE in cattle in the three countries where cases had occurred, the UK, France and more recently, the Republic of Ireland.

The Origin of Infection
The important question to answer was how did this infection arise? Could it have been due to high exposure to the BSE agent, which might be more likely in people working with high-risk tissues such as abattoir workers,

knackermen and veterinary pathologists? Or could it have been because high-risk tissues from cattle infected with the BSE agent were knowingly, or unknowingly, part of the human diet? MBM was not known to be part of the human diet. The alternative source, CNS tissue from cattle infected with the BSE agent, was a possibility, particularly historically.

By 1999, analyses of epidemiological data did not provide evidence to suggest that there is an increased risk of vCJD through past surgery, previous blood transfusion, occupation or a range of dietary factors [12]. Thus it is still unproved that cattle are the source of the infection the patients acquired, or if they are, that it was an exposure resulting from consumption of cattle tissues.

In March 1996, the risk from SBO was considered to be very low because of the effects of the SBO ban and its subsequent and recent modifications. Thus new exposures from this source would also be expected to be very low. However, at the time the SEAC announcement was made on 20 March 1996, the Committee stated that: 'Although there is no direct evidence of a link (*between BSE and vCJD*), on current data and in the absence of any credible alternative, the most likely explanation at present is that these (*ten*) cases are linked to exposure to BSE before the introduction of the SBO ban in 1989.' As a result of the new findings, the statement and the ensuing debate, the EC, at the behest of Member States of the EU, imposed a ban on exports from the UK of cattle and cattle products with the exception of milk and semen. The Commissioner made it clear that no action was to be taken at Commission level in regard to consumption of authorized beef products within the UK. That was the responsibility of the UK Government. The ban was for reasons of public confidence in the safety of beef. Subsequently, a set of conditions was agreed which the UK had to satisfy before beef could be exported again. The ban was lifted by Commission Decision in August 1999, challenged subsequently by France and the challenge rejected on scientific grounds by the EC Scientific Steering Committee on 29 October 1999.

Returning to the period following the discovery of vCJD, epidemiological, genetic, molecular and transmission experiments were set up or extended.

In regard to the molecular genetic investigations, results showed that there was no evidence for the existence of a *PrP* gene mutation in any confirmed case of vCJD. But each patient so far studied has exhibited methionine homozygosity at polymorphic codon 129 of this gene. This is a common polymorphism that occurs at a frequency of around 39% in the normal populations so far studied [112]. There was no evidence at all for the familial occurrence of vCJD.

Iatrogenic transmission of CJD was also ruled out [12].

Experimental Transmission

Transmission studies were critical. The objectives were to demonstrate whether or not vCJD was an experimentally transmissible disease, to identify an experimental host as a model for further studies and to define the agent biologically. Most important of all was to compare the strain type of agent isolated from human brain with that of the BSE agent derived from cattle brain, or with the strain type of any other agents that were relevant.

Transmission to Conventional Mice and Biological Strain Typing

Even before the experiments started, it was clear that the agents could not be precisely the same. This is because the amino acid sequence of the PrP from cattle was known to be different from that derived from humans. The significance of this fact depends on the viewpoint of the reader in regard to the nature of TSE agents i.e. whether they are prions, virinos or unconventional viruses.

Notwithstanding the different hypotheses of agent structure, it was already known from several previous transmissions to mice (from brains of cattle with BSE, cats with FSE, a nyala and greater kudu with TSE, and pigs sheep and goats with experimental BSE) that they all showed the same incubation period and lesion profile. That is to say the strain type isolated from each of these species was biologically indistinguishable. But it was different from any of the known 20 or so strains of scrapie agent [21, 72]. This was despite the fact that the amino acid sequence in PrP from each of these species was different. The authors also suggested that some additional informational molecule, such as a small as yet unidentified nucleic acid, might be required to define the strain. Since that time much work has been done to show how the PrP might confer strain-specific properties by virtue of its conformation [113–116]. However, the differing views about the agent structure have not yet been resolved.

Using the same kind of in-bred mice as referred to above, it was shown that isolates from brains from 2 patients each with sporadic CJD, the first 2 confirmed before the BSE era, 2 during the BSE era and 2 from farmers who had BSE in their herds, were all quite unlike the BSE agent [117]. However, when isolates from the brains of 3 patients with vCJD were inoculated and transmitted to the same kind of mice [117], 3 conclusions were drawn. First vCJD was an experimentally transmissible disease. Second the mice were uniformly susceptible. They were thus potentially useful as a model species to titrate the infectivity and investigate its distribution in body tissues. This was a matter of some importance considering the magnitude of surgical interventions in the general population and the potential risks that might ensue from them or other interventions, such as transfusion of blood derived from clinically healthy patients incubating the disease. Finally, it was conclusively shown that the strain of agent was effectively identical in three patients with vCJD and the strain was indistinguishable from the BSE strain from cattle.

Transmission to Transgenic Mice

A number of valuable studies using transgenic mice, commenced before vCJD was discovered, have helped to elucidate the strength of the species barrier between cattle and man with respect to the BSE agent. Mice that express both mouse and human PrP when challenged with CJD agent (excluding the vCJD strain that was not used) can generate human PrP^{Sc}. However when BSE agent is inoculated only mouse PrP^{Sc} is detected. This is not surprising since BSE from cattle from widely separated geographical locations readily and consistently transmits to conventional mice and is characterized by the deposition of PrP^{Sc} in brain tissue. The transgenic studies suggest that human PrP^{C} is less susceptible to being changed to human PrP^{Sc} by bovine prions than by human prions [118].

In a second study using transgenic mice that only expressed human PrP, the incubation period following challenge with CJD was shorter than when transgenic mice expressing mouse and human PrP were used, suggesting some degree of interference by the former, if PrP from both species was expressed. Furthermore infectious mouse brain material sub-passaged into the same kind of mice showed no change in incubation period, thus indicating that the species barrier had been abolished. When the challenge was with BSE agent to transgenic mice that only expressed human PrP, the incubation period was very long indeed. This suggests that human PrP^{C} is considerably more difficult to convert to PrP^{Sc} using bovine prions than human prions [118]. The fact that the mice used in the study expressed valine and not methionine at codon 129 means that predictions about the strength of the species barrier between cattle and mice cannot be firmly made. This is because cattle all express only methionine at this codon as do all human patients with vCJD so far tested.

CJD prions such as those used in the above experiments have been transmitted to wild-type mice in the past but often with some difficulty and sometimes not at all. However, vCJD prions transmit much more readily to the wild-type (FVB) mice used in the reported study, whereas transmission to transgenic mice is less efficient [119]. The outcome of a further series of transgenic mice studies, especially involving those with the codon 129 methionine homozygosity, is awaited with interest.

Molecular Strain Typing

Biological strain typing, generally regarded as the gold standard method for confirming identity between agents, is expensive, time consuming (at least 18 months to 2 years to get a result and sometimes longer) and of little practical value other than for research. Molecular strain typing [120], however, does not suffer from any of these criticisms and supports the view that the BSE agent from cattle could be responsible for vCJD [119, 121]. Assuming the prion hypothesis of agent structure is correct, strain variation is proposed to be dependent upon the conformation (shape and folding pattern) of PrP and the glycosylation ratios. Thus it is proposed that the PrP protein itself may encode the phenotype of the disease [121]. Molecular analysis of prion strain variation (molecular, as distinct from biological strain typing) is based on distinguishing between the biochemical properties of PrP derived from the brains of patients with different CJD phenotypes. Following extraction, treatment with proteinase K, Western blotting and determination of the glyco-sylation ratios in PrP^{Sc} derived from human brain, at least 4 types of PrP (types 1–4) can be identified [121]. There is a characteristic abundance of the diglycosylated form of PrP derived from the brain of vCJD cases. This profile has been seen also in transgenic mice to which the disease has been transmitted. Types 1–3 occurred in brains from sporadic, familial and iatrogenic CJD patients. Type 4 PrP was found only in the brains from patients with vCJD. It was found in all brains examined and in the tonsil too when it was sought. No PrP was found in the tonsils from patients with other forms of CJD. Thus there are possibilities to distinguish vCJD from other forms of CJD both in life (by tonsil biopsy) and after death [122].

Because the molecular strain type of agent identified in all cases of vCJD so far examined is consistent (type 4) and this type has not been seen in other forms of CJD, it is now regarded as a valuable aid to confirm the aetiological diagnosis. This is currently impractical by using mice for biological strain typing. Some authors nevertheless have aired a view of caution about the use of the method in its current form for determining the molecular strain types of agent isolated from other species [123–126]. This is because previously existing biologically typed strains can have molecular profiles similar to vCJD and BSE [123, 126]. Furthermore, PrP^{Sc} also varies independently in the amount and pattern of glycosylation according to brain region [125]. In addi-tion the molecular signature of PrP^{Sc} from 42 French isolates of natural scrapie from 21 flocks in different regions of France could not be distinguished from that in French cattle with BSE or from a cheetah imported from the UK with FSE [127]. There is no evidence for the occurrence of BSE in sheep in any country so it seems unwise at present to assume that molecular typing of isolates from different species is robust enough to certainly distinguish bio-logical strains. Technical differences between laboratories could account for some of the anomalous results so far reported [128, 129]. More technical development is required together with a harmonization of methods between laboratories.

Comparative Neuropathology

As mentioned in a previous section, BSE has been experimentally transmitted to macaque monkeys [79]. The clinical, and particularly the neuropathological features, resemble those seen in vCJD [79, 130]. However, these features have not been reported in 3 other species of primate challenged with BSE [78, C.J. Gibbs Jr., pers. commun.]. Furthermore there are similarities between the TSE observed in mule deer with chronic wasting disease and vCJD [4], particularly in the occurrence of PrP-positive plaques.

Tissue Distribution of Prion Protein – Scrapie Isoform

In regard to the tissue distribution of PrP^{Sc} in vCJD this has been found in the tonsil [122], appendix [131] spleen and lymph nodes [130]. Results from tissue infectivity studies have yet to be reported for tissues other than brain. Since PrP is not detected in the tonsils of patients with other forms of CJD, and infectivity or the presence of PrP^{Sc} is infrequently reported from LRS tissues of patients with other forms of CJD, it is possible that the pathogenesis of vCJD is different from the other forms too.

For a review of vCJD, see Collinge [131].

Effect of the Occurrence of New Variant Form of Creutzfeldt-Jakob Disease

The potential links of vCJD to the BSE agent from cattle via consumption of infected offals like brain and spinal cord before measures were put in place in 1989 have been sufficiently convincing to have caused severe repercussions on the beef industries of countries with BSE in native-born cattle. The occurrence of BSE in cattle in the UK alone has been estimated to have cost >£4 billion to 1999.

The proposed linkage between BSE and vCJD has resulted in the imposition of SBO bans, or amendment to offals bans in existence. In some countries, including the UK, the SBO ban was extended to a specified bovine materials ban and thereafter to a specified risk materials ban that included some sheep tissues. An EC Decision on the exclusion of specified risk materials has been drafted but not yet agreed by EU Member States.

The UK ban included the central nervous system of sheep and goats of certain ages and the spleen of sheep and goats of all ages, as a hypothetical risk from BSE in sheep was perceived. This was because some sheep had been exposed to MBM in their diet before the feed ban (and perhaps by accident afterwards) and BSE had been experimentally transmitted to both sheep and goats by the oral route [80]. No natural case of BSE has been found in sheep anywhere in the world, even after limited investigations using biological strain typing of isolates or molecular typing or PrP from sheep [133].

The potential for sheep to harbour the BSE agent and for it to be transmitted maternally and horizontally has been one of the most difficult to come to terms with. Adopting the precautionary principle and excluding all at risk sheep and even tissues from healthy sheep would have very serious consequences for the sheep industry of Europe and especially the UK and its environment. Sheep play a major role in the rural scene and the consequences from a socio-economic point of view are incalculable. On the other hand doing nothing could put consumers at risk from exposure to BSE agent if it was in fact present and if it is pathogenic for man. At present, some countries in the EU, including the UK, have opted for a sensible risk reduction policy. This may have to be modified if ever BSE was found in sheep. Humans have been widely exposed to sheep PrP over centuries without any evidence of risk for human health. We cannot be totally confident that that could not happen in the future as a result of exposure to the new (currently potential) hazard of BSE in sheep. However, as we acquire new knowledge of prion diseases and how they work, it seems that it may become possible to control, and eventually eliminate, clinical scrapie from sheep. How that might be done and whether infection will be eliminated too is another story beyond the compass of this chapter.

Concluding Remarks

BSE is well on the way to eradication in the UK, albeit with some hiccups on the way. The UK experience of BSE shows how, with perseverance and using multidisciplinary techniques, a serious epidemic of a new unconventional disease can be first brought under control and then eliminated. However, there are many lessons to be learned and one of them is that it should never have been allowed to happen in the first place.

Acknowledgments

The author thanks many colleagues at the Institute for Animal Health including Dr. M.E. Bruce, Mr. J.D. Foster, Dr. H. Fraser and Dr. D.M. Taylor; Mr. S.A.C. Hawkins, Mr. G.A.H. Wells, and Mr. J.W. Wilesmith from the Veterinary Laboratories Agency, and Dr. D. Matthews of MAFF Tolworth, all for access to unpublished data and the library staff at the Veterinary Laboratory Agency for provision of published material.

References

1 Wells GAH, Scott AC, Johnson CT, Gunning RF, Hancock RD, Jeffrey M, Dawson M, Bradley R: A novel progressive spongiform encephalopathy in cattle. Vet Rec 1987;121:419–420.
2 Parry HB: Scrapie disease in sheep. Oppenheimer DR (ed). London, Academic Press, 1983, p 192.
3 Marsh RF, Hadlow WJ: Transmissible mink encephalopathy. Rev Sci Tech 1992;11:539–550.
4 Williams ES, Young S: Spongiform encephalopathies in Cervidae. Rev Sci Tech 1992;11:551–567.
5 Gajdusek DC, Zigas V: Degenerative disease of the central nervous system in New Guinea – The endemic occurrence of 'kuru' in the native population. N Engl J Med 1975;257:974–978.
6 Anon: Report of the Chief Veterinary Officer. London, HMSO; Animal Health, 1986, p 69.
7 MAFF, 1999. Bovine spongiform encephalopathy in Great Britain. A Progress Report, Tolworth, MAFF, June 1999.
8 Hope J, Reekie LJD, Hunter N, Multhaup G, Beyreuther K, White H, Scott AC, Stack MJ, Dawson M, Wells GAH: Fibrils from brains of cows with new cattle disease contain scrapie-associated protein. Nature 1988;336:390–392.
9 Korth C, Stierli B, Streit P, Moser M, Schaller O, Fischer R, Schulz-Schaeffer W, Kretzschmar H, Raeber A, Braun U, Ehrensperger F, Hornemann S, Glockshuber R, Riek R, Billeter M, Wüthrich K, Oesch B: Prion (PrP^{Sc})-specific epitope defined by a monoclonal antibody. Nature 1997;390:74–77.
10 Somerville RA, Dunn AJ: The association between PrP and infectivity in scrapie and BSE infected mouse brain. Arch Virol 1996;141:275–289.
11 Will RG, Ironside JW, Zeidler M, Cousens SN, Estibeiro K, Alperovitch A, Poser S, Pocchiari M, Hofman A, Smith PG: A new variant of Creutzfeldt-Jakob disease in the UK. Lancet 1996;347:921–925.
12 CJD Report 1998: Creutzfeldt-Jakob disease in the UK. London, Department of Health; p 51.
13 Ghani AC, Ferguson NM, Donnelly CA, Hagenaars TJ, Anderson RM: Epidemiological determinants of the pattern and magnitude of the vCJD epidemic in Great Britain. Proc R Soc Lond B 1998;265:2443–2452.
14 Department of Health 1999/0646. Monthly Creutzfeldt-Jakob disease figures. London, Department of Health; Press Release, 1 November 1999.
15 Schreuder BEC: BSE agent hypotheses. Livestock Prod Sci 1994;38:23–33.
16 Taylor DM: Exposure to, and inactivation of, the unconventional agents that cause transmissible degenerative encephalopathies; in Baker H, Ridley RM (eds): Methods in Molecular Medicine. Totowa, Humana Press, 1996, pp 105–118.
17 Garland AJM: A review of BSE and its inactivation. Eur J Parent Sci 1999;4:86–93.
18 Prusiner SB: Novel proteinaceous particles cause scrapie. Science 1982;216:136–144.
19 Aguzzi A, Brandner S: Shrinking prions: New folds to old questions. Nat Med 1999;5:486–487.
20 Bruce ME: Bovine spongiform encephalopathy: Experimental studies. OIE/WHO Consultation on BSE. Paris, OIE, 1994.
21 Bruce ME: Strain typing studies of scrapie and BSE; in Baker H, Ridley RM (eds): Methods in Molecular Medicine: Prion Diseases. Totowa, Humana Press, 1996, pp 223–236.
22 Goldmann W, Hunter N, Martin T, Dawson M, Hope J: Different forms of the bovine PrP gene have five or six copies of a short, G-C-rich element within the protein-coding exon. J Gen Virol 1991;72:201–204.
23 McKenzie DI, Cowan CM, Marsh RF, Aiken JM: PrP gene variability in the US cattle population. Anim Biotechnol 1992;3:309–315.
24 Grobet L, Vandevenne S, Charlier C, Pastoret PP, Hanset R: Polymorphisme du gène de la protéine prion chez des bovins belges. Ann Méd Vét 1994;138:581–586.
25 Hunter N, Goldmann W, Smith G, Hope, J: Frequencies of PrP gene variants in healthy cattle and cattle with BSE in Scotland. Vet Rec 1994;135:400–403.
26 Wilesmith JW, Wells GAH: Bovine spongiform encephalopathy. Curr Topics Microbiol Immunol 1991;172:21–38.

27 Wilesmith JW, Hoinville LJ, Ryan JBM, Sayers AR: Bovine spongiform encephalopathy: Aspects of the clinical picture and analyses of possible changes 1986–1990. Vet Rec 1992;130:197–201.

28 Braun U, Schicker E, Hürnlimann B: Diagnostic reliability of clinical signs in cows with suspected bovine spongiform encephalopathy. Vet Rec 1998;143:101–105.

29 Andrews T: The 'downer cow'. Practice 1986;September:187–189.

30 Weaver AD: Bovine spongiform encephalopathy: Its clinical features and epidemiology in the United Kingdom and significance for the United States. CompendFood Animal 1992;14:1647–1655.

31 Hoinville LJ: Clinical signs of reported cases of BSE and their analysis to aid in diagnosis. Cattle Pract 1993;1:59–62.

32 Austin AR, Simmons MM: Reduced rumination in bovine spongiform encephalopathy and scrapie. Vet Rec 1993;132:324–325.

33 Aldridge BM, Scott PR, Clarke M, Will R, McInnes A: Proc 15th World Buiatrics Congress, Riva del Gardi, 1998, p 1531.

34 Winter MH, Aldridge BM, Scott PR, Clarke M: Occurrence of 14 cases of bovine spongiform encephalopathy in a closed dairy herd. Br Vet J 1989;145:191–194.

35 Austin AR, Pawson L, Meek S, Webster S: Abnormalities of heart rate and rhythm in bovine spongiform encephalopathy. Vet Rec 1997;141:352–357.

36 Wilesmith JW, Wells GAH, Cranwell MP, Ryan JBM: Bovine spongiform encephalopathy: Epidemiological studies. Vet Rec 1988;123:638–644.

37 Wilesmith JW, Ryan JBM, Atkinson MJ: Bovine spongiform encephalopathy: Epidemiological studies on the origin. Vet Rec 1991;128:199–203.

38 Wilesmith JW, Ryan JBM, Hueston WD: Bovine spongiform encephalopathy: Case-control studies of calf feeding practices and meat and bone meal inclusion in proprietary concentrates. Res Vet Sci 1992;52:325–331.

39 Fraser H, McConnell I, Wells GAH, Dawson M: Transmission of bovine spongiform encephalopathy to mice. Vet Rec 1988;123:472.

40 Order. Supplies and services (feeding stuffs) (GB). The feeding stuffs (manufacture) Order 1950. Statutory Instrument 1950: No 1988. London, HMSO.

41 Richards MS, Wilesmith JW, Ryan JBM, Mitchell AP, Wooldridge MJA, Sayers AR, Hoinville LJ: Methods of predicting BSE incidence; in Thrusfield MV (ed): Proc Soc Vet Epidemiol Prev Med, Exeter 1993. Edinburgh, The Society, 1993, pp 70–81.

42 Taylor DM, Fernie K, McConnell I, Ferguson CE, Steele PJ: Solvent extraction as an adjunct to rendering: The effect on BSE and scrapie agents of hot solvents followed by dry heat and steam. Vet Rec 1998;143:6–9.

43 Taylor DM, Woodgate SL, Atkinson MJ: Inactivation of the bovine spongiform encephalopathy agent by rendering procedures. Vet Rec 1995;137:605–610.

44 Taylor DM, Woodgate SL, Fleetwood AJ, Cawthorne RJG: Effect of rendering procedures on the scrapie agent. Vet Rec 1997;141:643–649.

45 Commission Decision 96/449/EC: On the approval of alternative heat treatment systems for processing animal waste with a view to inactivation of spongiform encephalopathy agents. Off J Euro Comm 1996; L 184/43.

46 Hau CM, Curnow RN: Separating the environmental and genetic factors that may be causes of bovine spongiform encephalopathy. Phil Trans R Soc Lond B 1996;351:913–920.

47 Bradley R, Wilesmith JW: Epidemiology and control of bovine spongiform encephalopathy (BSE). Br Med Bull 1993;49:932–959.

48 Dawson M, Wells GAH, Parker BNJ: Preliminary evidence of the experimental transmissibility of bovine spongiform encephalopathy to cattle. Vet Rec 1990;126:112–113.

49 Anderson RM, Donnelly CA, Ferguson NM, Woolhouse MEJ, Watt CJ, Mawhinney S, Dunstan SP, Southwood TRE, Wilesmith JW, Ryan JBM, Hoinville LJ, Hillerton JE, Austin AR, Wells GAH: Transmission dynamics and epidemiology of BSE in British cattle. Nature 1996;382:779–788.

50 Wilesmith JW: Manual on bovine spongiform encephalopathy. Rome, FAO: 1998, p 51.

51 Wilesmith JW: Bovine spongiform encephalopathy: Epidemiological factors associated with the emergence of an important new animal pathogen in Great Britain. Semin Virol 1994;5:179–187.

52 Kimberlin RH: A scientific evaluation of research into bovine spongiform encephalopathy (BSE); in Bradley R, Marchant B (eds): Transmissible Spongiform Encephalopathies. Proceedings of a consultation with the Scientific Veterinary Committee of the CEC 14–15 September 1993. VI/4131/94-EN Brussels EC 1994; pp 455–477.

53 Kimberlin RH, Wilesmith JW: Bovine spongiform encephalopathy epidemiology, low dose expose and risks. Ann N Y Acad Sci 1994;724:210–220.

54 Cutlip RC, Miller JM, Race RE, Jenny AL, Katz JB, Lehmkuhl HD, DeBey BM, Robinson MM: Intracerebral transmission of scrapie to cattle. J Infect Dis 1994;169:814–820.

55 Hill F: Neurological diseases of cattle where BSE has been included in the differential diagnosis. Surveillance 1994;21:25.

56 Lloyd-Webb E: BSE awareness programme in Tasmania. Vet Rec 1994;134:480.

57 Anon: A retrospective study of the nervous diseases of sheep and cattle by the Official Laboratory and School of Veterinary Medicine from 1972–1996. Internal Report 1997 Montevideo, Uruguay.

58 Kimberlin RH, Cole S, Walker CA: Temporary and permanent modifications to a single strain of mouse scrapie on transmission to rats and hamsters. J Gen Virol 1987;68:1875–1881.

59 Hoinville LJ, Wilesmith JW, Richards MS: An investigation of risk factors for cases of bovine spongiform encephalopathy born after the introduction of the 'feed ban'. Vet Rec 1995;136:312–318.

60 Wilesmith JW, Wells GAH, Ryan JBM, Gavier-Widen D, Simmons MM: A cohort study to examine maternally-associated risk factors for bovine spongiform encephalopathy. Vet Rec 1997;141:239–243.

61 Curnow RN, Hau CM: The incidence of bovine spongiform encephalopathy in the progeny of affected sires and dams. Vet Rec 1996;138:407–408.

62 Donnelly CA, Gore SM, Curnow RN, Wilesmith JW: The Bovine Encephalopathy Maternal Cohort Study: Its purpose and findings. Appl Statist 1997;46:299–304.

63 Donnelly CA, Ghani AC, Ferguson NM, Wilesmith JW, Anderson RM: Analysis of the Bovine Spongiform Encephalopathy Maternal Cohort Study: Evidence for direct maternal transmission. Appl Statist 1997;46:321–344.

64 Spongiform Encephalopathy Advisory Committee (SEAC), Annual Report 1997/98; PB4157; MAFF, London DoH, 1998.

65 Scientific Steering Committee: Opinion on the possible vertical transmission of bovine spongiform encephalopathy (BSE). Adopted 18–19 March 1999. Brussels, EC, 1999.

66 Moynagh J, Schimmel H: Tests for BSE evaluated. Nature 1999;400:105.

67 Wyatt JM, Pearson GR, Smerdon T, Gruffydd-Jones TJ, Wells GAH: Spongiform encephalopathy in a cat. Vet Rec 1990;126:513.

68 Pearson GR, Wyatt JM, Henderson JP, Gruffydd-Jones TJ: Feline spongiform encephalopathy: A review. Vet A 1993;33:1–10.

69 Kirkwood JK, Cunningham AA: Epidemiological observations on spongiform encephalopathies in captive wild animals in the British Isles. Vet Rec 1994;135:296–303.

70 Bratberg B, Ueland K, Wells GAH: Feline spongiform encephalopathy in a cat in Norway. Vet Rec 1995;136:444.

71 Zanusso G, Nardelli E, Rosati A, Fabrizi G-M, Ferrari S, Carteri A, De Simone F, Rizzuto N, Monaco S: Simultaneous occurrence of spongiform encephalopathy in a man and his cat in Italy. Lancet 1998;352:1116–1117.

72 Bruce ME, Chree A, McConnell I, Foster J, Pearson G, Fraser H: Transmission of bovine spongiform encephalopathy and scrapie to mice: Strain variation and the species barrier. Phil Trans R Soc Lond B 1994;343:405–411.

73 Schoon HA, Brunckhorst D, Pohlenz J: Spongiform encephalopathy in a red-necked ostrich (*Struthio camelus*), a case history. Tierärztl Praxis 1991;19:263–265.

74 Schoon HA, Brunckhorst D, Pohlenz JA: A contribution to the neuropathlolgy of the red-necked ostrich (*Struthio camelus*) – Spongiform encephalopathy. VerhBber Erkr Zootiere 1991;33:309–314.

75 Bons N, Mestre-Frances N, Charnay Y, Salmona M, Tagliavini F: Encéphalopathie spongiforme spontanée chez un jeune singe rhésus adulte. Sci Méd 1996;319:733–736.

76 Bons N, Mestre-Frances N, Belli P, Cathala F, Gajdusek DC, Brown P: Naturaland experimental oral infection of nonhuman primates by bovine spongiform encephalopathy agents. Proc Natl Acad Sci USA 1999;96:4046–4051.

77 Ridley RM, Baker HF: Oral transmission of BSE to primates. Lancet 1996; 348:1174.

78 Baker HF, Ridley RM, Wells GAH: Experimental transmission of BSE and scrapie to the common marmoset. Vet Rec 1993;132:403–406.

79 Lasmézas CI, Deslys J-P, Demalmay R, Adjou KT, Lamoury F, Dormont D, Robain O, Ironside J, Hauw J-J: BSE transmission to macaques. Nature 1996;381:743–744.

80 Foster JD, Hope J, Fraser H: Transmission of bovine spongiform encephalopathy to sheep and goats. Vet Rec 1993;133:339–341.

81 Dawson M, Wells GAH, Parker BNJ, Scott AC: Primary parenteral transmission of bovine spongiform encephalopathy to the pig. Vet Rec 1990;127:338.

82 Robinson MM, Hadlow WJ, Huff TP, Wells GAH, Dawson M, Marsh RF, Gorham JR: Experimental infection of mink with bovine spongiform encephalopathy. J Gen Virol 1994;75:2151–2155.

83 Fraser H, Foster JD: Transmission to mice, sheep and goats and bioassay of bovine tissues; in Bradley R, Marchant B (eds): Transmissible Spongiform Encephalopathies. Proceedings of a consultation with the Scientific Veterinary Committee of the CEC 14–15 September 1993. VI/4131/94–EN Brussels, EC, 1994, pp 145–159.

84 Taylor DM, Ferguson CE, Bostock CJ, Dawson M: Absence of disease in mice receiving milk from cows with bovine spongiform encephalopathy. Vet Rec 1995;136:592.

85 Middleton DJ, Barlow RM: Failure to transmit bovine spongiform encephalopathy to mice by feeding them with extraneural tissues of affected cattle. Vet Rec 1993;132:545–547.

86 Wrathall AE: Risks of transmitting scrapie and bovine spongiform encephalopathy by semen and embryos. Rev Sci Tech 1997;16:240–264.

87 Taylor DM, Ferguson CE, Chree A: Absence of detectable infectivity in trachea of BSE-affected cattle. Vet Rec 1996;138:160–161.

88 Fraser H, Bruce ME, Chree A, McConnell I, Wells GAH: Transmission of bovine spongiform encephalopathy to mice. J Gen Virol 1992;73:1891–1897.

89 Taylor DM, Brown JM, Fernie K, McConnell I: The effect of formic acid on BSE and scrapie infectivity in fixed and unfixed brain-tissue. Vet Microbiol 1997;58:167–174.

90 Kimberlin RH: Early events in the pathogenesis of scrapie in mice: Biological and biochemical studies; in Prusiner SB, Hadlow WJ (eds): Slow Transmissible Diseases of the Nervous System. New York, Academic Press, 1979, vol 2, pp 33–54.

91 Klein MA, Frigg R, Raeber AJ, Flechsig E, Hegyi I, Zinkernagel RM, Weissmann C, Aguzzi A: PrP expression in B lymphocytes is not required for prion neuroinvasion. Nat Med 1998;4:1429–1433.

92 Kimberlin RH, Walker CA: Pathogenesis of experimental scrapie; in Bock G, Marsh J (eds): Novel Infectious Agents and the Central Nervous System. Ciba Foundation Symposium; Wiley, Chichester, 1988, vol 135, pp 37–62.

93 Beekes ME, Baldauf E, Diringer H: Sequential appearance and accumulation of pathognomonic markers in the central nervous system of hamsters orally infected with scrapie. J Gen Virol 1996; 77:1925–1934.

94 Baldauf E, Beekes M, Diringer H: Evidence for an alternative direct route of access for the scrapie agent to the brain bypassing the spinal cord. J Gen Virol 1997;78:1187–1197.

95 Beekes M, McBride P, Baldauf E: Cerebral targeting indicates vagal spread of infection in hamsters fed with scrapie. J Gen Virol 1998;79:601–607.

96 McBride PA, Eikelenboom P, Kraal G, Fraser H, Bruce ME: PrP protein is associated with follicular dendritic cells of spleens and lymph nodes in uninfected mice and scrapie-infected mice. J Pathol 1992;168:413–418.

97 Berg LJ: Insights into the role of the immune system in prion diseases. Proc Natl Acad Sci USA 1994;91:429–432.

98 Aguzzi A: Pathogenesis of spongiform encephalopathies: An update. Int Arch Allergy Immunol 1996;110:99–106.

99 Hadlow WJ, Kennedy RC, Race RE, Eklund CM: Virological and neurohistological findings in dairy goats affected with natural scrapie. Vet Pathol 1980;17:187–199.

100 Hadlow WJ, Kennedy RC, Race RE: Natural infection of Suffolk sheep with scrapie virus. J Infect Dis 1982;146:657–664.

101 Hadlow WJ, Race RE, Kennedy RC, Eklund CM: Natural infection of the sheep with scrapie virus; in Prusiner SB, Hadlow WJ (eds): Slow Transmissible Diseases of the Nervous System. New York, Academic Press, 1979, vol 2, pp 3–12.

102 Hoinville LJ: A review of the epidemiology of scrapie in sheep. Rev Sci Tech 1996;15:827–852.

103 Pattison IH, Hoare MN, Jebett JN, Watson WA: Spread of scrapie to sheep and goats by oral dosing with foetal membranes from scrapie affected sheep. Vet Rec 1972;90:465–468.

104 Pattison IH, Hoare MN, Jebett JN, Watson WA: Further observations on the production of scrapie on sheep by oral dosing with foetal membranes from scrapie affected sheep. Br Vet J 1974;130:lxv–lxvii.

105 Onodera T, Ikeda T, Muramatsu Y, Shinagawa M: Isolation of scrapie agent from the placenta of sheep with natural scrapie in Japan. Microbiol Immunol 1993;37:311–316.

106 Race RE, Jenny A, Sutton D: Scrapie infectivity and proteinase K-resistant prion protein in sheep placenta, brain, spleen, and lymph node: Implications for transmission and ante mortem diagnosis. J Infect Dis 1998;178:949–953.

107 Foster JD, Bruce M, McConnell I, Chree A, Fraser H: Detection of BSE infectivity in brain and spleen of experimentally infected sheep. Vet Rec 1996;138:546–548.

108 Wells GAH, Dawson M, Hawkins SAC, Green RB, Dexter I, Francis ME, Simmons MM, Austin AR, Horigan MW: Infectivity in the ileum of cattle challenged orally with bovine spongiform encephalopathy. Vet Rec 1994;135:40–41.

109 Wells GAH, Dawson M, Hawkins SAC, Austin R, Green RB, Dexter I, Horigan MW, Simmons MM: Preliminary observations on the pathogenesis of experimental bovine spongiform encephalopathy; in Gibbs CJ Jr (ed): Bovine Spongiform Encephalopathy. The BSE Dilemma. New York, Springer, 1996, pp 28–44.

110 Wells GAH, Hawkins SAC, Green RB, Austin AR, Dexter I, Spencer YI, Chaplin MJ, Stack MJ, Dawson M: Preliminary observations on the pathogenesis of experimental bovine spongiform encephalopathy (BSE): An update. Vet Rec 1998;142:103–106.

111 Wells GAH, Hawkins SAC, Green RB, Spencer YI, Dexter I, Dawson M: Limited detection of sternal bone marrow infectivity in the clinical phase of experimental bovine spongiform encephalopathy (BSE). Vet Rec 1999;144:292–294.

112 Alperovitch A, Zerr I, Pocchiari M, Mitrova E, de Pedro Cuesta J, Hegyi I, Collins S, Kretzschmar H, van Duijn C, Will RG: Codon 129 prion protein genotype and sporadic Creutzfeldt Jakob disease. Lancet 1999;353:1673–1674.

113 Telling GC, Parchi P, DeArmond SJ, Cortelli P, Montagna P, Gabizon R, Mastrianni J, Lugaresi E, Gambetti P, Prusiner SB: Evidence for the conformation of the pathologic isoform of the prion protein enciphering and propagating prion diversity. Science 1996;274:2079–2082.

114 Safar J, Wille H, Itri V, Groth D, Serban H, Torchia M, Cohen FE, Prusiner SB: Eight prion strains have PrP^{Sc} molecules with different conformations. Nat Med 1998;4:1157–1165.

115 Aguzzi A: Protein conformation dictates prion strain. Nat Med 1998;4:1125–1126.

116 Wadsworth JDF, Jackson GS, Hill AF, Collinge J: Molecular biology of prion propagation. Curr Opin Gen Dev 1999;9:338–345.

117 Bruce ME, Will RG, Ironside JW, McConnell I, Drummond D, Suttle A, McCardle L, Chree A, Hope J, Birkett C, Cousens S, Fraser H, Bostock CJ: Transmissions to mice indicate that 'new variant' CJD is caused by the BSE agent. Nature 1997;389:498–501.

118 Collinge J, Palmer MS, Sidle KCL, Hill AF, Gowland I, Meads J, Asante E, Bradley R, Doey LJ, Lantos PL: Unaltered susceptibility to BSE in transgenic mice expressing human prion protein. Nature 1995;378:779–783.

119 Hill AF, Desbruslais M, Joiner S, Sidle KCL, Gowland I, Collinge J, Doey LJ, Lantos P: The same prion strain causes vCJD and BSE. Nature 1997;389:448–450.

120 Parchi P, Castellani R, Capellari S, Ghetti B, Young K, Chen SG, Farlow M, Dickson DW, Sima AAF, Trojanowski JQ, Petersen RB, Gambetti P: Molecular basis of phenotypic variability in sporadic Creutzfeldt-Jakob disease. Ann Neurol 1996;39:669–680.

121 Collinge J, Sidle KCL, Meads J, Ironside J, Hill AF: Molecular analysis of prion strain variation and the aetiology of 'new variant' CJD. Nature 1996;383:685–690.

122 Hill AF, Zeidler M, Ironside J, Collinge J: Diagnosis of new variant Creutzfeldt-Jakob disease by tonsil biopsy. Lancet 1997;349:99–100.

123 Somerville RA, Chong A, Mulqueen OU, Birkett CR, Wood SCER, Hope J: Biochemical typing of scrapie strains. Nature 1997;386:564.
124 Collinge J, Hill AF, Sidle KCL: Biochemical typing of scrapie strains. Nature 1997;386:564.
125 Somerville RA: Host and transmissible spongiform encephalopathy agent strain control glycosylation of PrP. J Gen Virol 1999;80:1865–1872.
126 Hope J, Wood SCER, Birkett CR, Chong A, Bruce ME, Cairns D, Goldmann W, Hunter N, Bostock CJ: Molecular analysis of ovine prion protein identifies similarities between BSE and an experimental isolate of natural scrapie, CH1641. J Gen Virol 1999;80:1–4.
127 Baron TGM, Madec J-Y, Calavas D: Similar signature of the prion protein in natural sheep scrapie and bovine spongiform encephalopathy-linked diseases. J Clin Microbiol 1999;37:3701–3704.
128 Parchi P, Capellari S, Chen SG, Petersen RB, Gambetti P, Kopp N, Brown P, Kitamoto T, Tateishi J, Giese A, Kretzschmar H: Typing prion isoforms. Nature 1997;386:232–233.
129 Collinge J, Hill AF, Sidle KCL, Ironside J: Typing prion isoforms. Nature 1997;386:233–234.
130 Ironside JW: nvCJD: Exploring the limits of our understanding. Biologist 1999;46:172–176.
131 Collinge J: Variant Creutzfeldt-Jakob disease. Lancet 1999;354:317–323.
132 Hilton A, Fathers E, Edwards P, Ironside JW, Zajicek J: Prion immunoreactivity in appendix before clinical onset of variant Creutzfeldt-Jakob disease. Lancet 1998;352:703–704.
133 Hill AF, Sidle KCL, Joiner S, Keyes P, Martin TC, Dawson M, Collinge J: Molecular screening of sheep for bovine spongiform encephalopathy. Neurosci Lett 1998;255:159–162.

Mr. R. Bradley, contact address: Veterinary Laboratories Agency,
New Haw, Addlestone, KT15 3NB (UK)
Tel. +44 1932 357306, Fax +44 1932 354929, E-Mail raybradley@btinternet.com

Rabenau HF, Cinatl J, Doerr HW (eds): Prions. A Challenge for Science,
Medicine and Public Health System. Contrib Microb. Basel, Karger, 2001, vol 7, pp 145–150

....

Animal Transmissible Spongiform Encephalopathy: Clinical and Diagnostic Aspects

Oskar-Rüger Kaaden

Institute for Medical Microbiology, Infectious and Epidemic Diseases,
Ludwig Maximilian University, Munich, Germany

Scrapie (German: Traberkrankheit; French: la tremblante du mouton; Islandic: Rida) is the prototype disease and the best-studied model of transmissible spongiform encephalopathies (TSEs) of humans and animals [1–4]. The TSE complex in humans comprises: kuru, Creutzfeldt-Jakob disease (CJD), Gerstmann- Sträussler-Scheinker (GSS) syndrome and fatal familial insomnia (FFI) [5, 6]. In addition to scrapie, other TSEs in animals are transmissible mink (*Mustela vison*) encephalopathy (TME) and chronic wasting disease of captive mule, deer and elk, and feline spongiform encephalopathy (FSE). For further details about TSE diseases in animals, see table 1 and also Bradley [7, this issue]. Swine could be infected after parenteral infection with bovine-spongiform-encephalopathy (BSE)-containing material, but not by natural infection.

Scrapie in sheep and, rarely in goats, has been known for more than 200 years and is endemic in many parts of the world. There are a few notable exceptions, like Australia and New Zealand, which are considered to be free of scrapie.

The disease is characterized by a long incubation period, in general about 3–4 years. The lack of an intra vitam diagnostic test makes it difficult to ascertain the existence of scrapie. As a matter of fact, it is impossible to explain the existence of scrapie on the basis of clinical surveillance. However, there is a genetic control of the incubation period which relies on isolate differences and genotype of the host [8, 9].

It is the only TSE which occurs as an endemic infection in its natural host, namely sheep and goats [10].

Table 1. TSEs in animals and humans

Disease	Species
Scrapie	sheep, goat
BSE	cattle, wild ruminants
FSE	Felidae
Chronic wasting disease of captive mule, deer and elk	wild ruminants
TME	mink
CJD	humans
vCJD	
GSS	humans
FFI	humans
Kuru	humans

Aetiology

Scrapie is an infectious disease which is mainly horizontally transmitted. A vertical mode of transmission has also been described [2]. The nature of the agent, however, is still in dispute. In recent years, the most widely accepted theory is that an infectious protein designated PrP^{Sc} (proteinase-resistant protein [5]; 'Sc' for scrapie) is responsible for the infectivity and thus for the disease. These structures called 'scrapie-associated fibrils', can be isolated from the brains of diseased sheep by biochemical methods and demonstrated by electron microscopy. This may occur as an autocatalytic enzymatic process between physiological and pathological molecules of PrP, an explanation which would also help to explain the long incubation time between the uptake of the infectious material and the outbreak of clinical scrapie-specific symptoms.

Pathogenesis

The basic mechanisms of the pathogenesis of scrapie are not well understood. Epidemiological observations suggest that the oro-pharynx and intestine are the main ports of entry of the infectious agent and the sites where its replication begins. The source of the infectious agent is not known since all examined excretions and secretions, like saliva, milk, urine and faeces, were free of detectable infectivity. However, the agent was regularly found in the intestinal mucosa and often in the mucosa of the upper respiratory tract in scrapie sheep. It is believed that the uptake of placental membranes from aborted fetuses from scrapie-infected sheep on pastures is an important route of horizontal transmission.

Sporadic reports mention that the scrapie could be diagnosed in various parts of the reproductive tract (ovary and uterus), placenta and aborted fetuses of infected ewes. There are no indications that rams play a major role in the transmission of the disease, as all attempts to detect infectivity in semen, testis or seminal vesicles have failed so far.

The present findings indicate that the early replication of the infectious scrapie agent occurs in the intestinal and lymphoreticular tissues, spleen, retropharyngeal and mesenteric lymph nodes. The spleen seems to play an important role in the pathogenesis of scrapie, as its surgical removal prolongs the incubation period and the onset of clinical symptoms.

Epidemiology

Scrapie has been known as an infectious disease in sheep for about 200 years. Yet, the mode of transmission and the prevalence of the disease are hardly understood. The most important point is that an intra vitam test is not possible. Therefore, although scrapie is a notifiable disease, it can only be diagnosed at autopsy and thus the published data are incomplete and do not reflect the real epidemiological situation. Scrapie is endemic in some European countries. In Germany, only sporadic clinical outbreaks have occurred during the last decade.

Clinical Symptoms

Scrapie is a neurodegenerative and non-febrile disorder in sheep, sometimes also in goats. Due to the degeneration of the nerve cells, affected animals show three different clinical symptoms: behavioural changes, alterations in the locomotor coordination and sensory behaviour. A typical feature of the disease is pruritus. This sign of 'scraping' or rubbing leads to loss of wool. The outcome of the disease is always fatal within 1–6 months.

Disease onset is generally marked by behavioural changes. Affected animal become nervous, sometimes even aggressive and isolate themselves from the flock. Hypersensitivity is another typical characteristic of scrapie. The animals develop tremor up to convulsion-like episodes. There is a decrease in body weight although the appetite seems to be retained.

The motor abnormalities are characterized by a high-stepping gait ('trotting') of the forelimbs. The movement of the hindlegs is described as a 'bunny hop'.

In the final stage, the scrapie-affected sheep develop severe ataxia and die within a few days. There are differences in the clinical symptoms depending on the scrapie strain involved and the genotype of the host.

Scrapie can be experimentally transmitted to mice and hamsters. The pathohistological findings broadly resemble those found in sheep and goats.

Although scrapie is one of the best-studied animal infectious diseases, some important questions are still open: Which is the minimal infectious dose for sheep? By which route and at what time is the infectious agent spread? Is the spreading of the PrP^{Sc} continuous or intermittent?

Pathological and Histopathological Lesions

The major histopathological finding is vacuolation of the nerve and glial cells, predominantly in the mesencephalon, medulla oblongata and pons. The degenerative processes lead to single and multiple vacuoles and finally to a spongiform appearance in the affected CNS. The severity of the clinical manifestations correlates well with the histological lesions found in the cerebellum. The way in which the scrapie agent reaches the CNS is still under investigation. Recent studies suggest that lymphocytes, via the peripheral blood and spleen and intra-axonal replication, may contribute to the localized expression of the scrapie-specific symptoms. All other internal organs are not affected by specific alterations.

It is well established by experimental studies in sheep that various scrapie strains and the host genetics affect the pathogenesis and finally the outcome. With regard to genetics, sheep have a major gene (*SiP*) with 2 alleles sA and pA which determines the incubation period of experimental scrapie and of natural scrapie in some sheep breeds.

Laboratory Diagnosis

As already mentioned, there is no reliable in vitram laboratory diagnosis for scrapie. PrP^{Sc}, as an autologous specific structural protein, does not stimulate the immune response. In addition, there is a lack of convenient methods for the detection and assay of the scrapie agent. Although some cell cultures support the replication of the agent, these methods are not reliable and practicable. In vivo investigations using intracerebroventricularly inoculated mice or hamsters are work-intensive and time consuming as well. Therefore, the histological detection of spongiform encephalopathy in stain brains section remains the method of choice. In addition, the isolation of PrP fibriles followed by immunoblotting may be used as molecular biological techniques in specialized laboratories.

The standard diagnostic method is the pathohistolological examination of brain sections.

Eradication

Scrapie is a notifiable disease in all member states of the European Union. The positive histological diagnosis results in the culling of the sheep flock. The carcasses have to be autoclaved according to the legal regulations at 132–135 °C for at least 20 min. Neither prophylactic vaccination programmes nor intra vitam diagnostic tests are available. During recent years, only a few sporadic cases of scrapie occurred in Germany. All affected herds were destroyed to prevent further transmission of the disease.

The scrapie agent, like other TSE agents, shows an extremely high degree of resistance [11] against disinfectants. Under experimental conditions with scrapie-infected syrian hamster brain (titre $10^{7.1}$ ID_{50}), autoclaving at 135 °C for 60 min was necessary to inactivate the infectivity completely. In field experiments, contaminated soil still contained infectivity and there was only a loss of 2–3 logs after 3 years.

Infectivity has also been found in some nematodes, like *Haemonchus contortus*, *Nippostrongylus brasiliensis* and *Syphacia obvoleta*. However, transmission experiments with these parasites were not successful.

Possible Zoonotic Character and Relation to Bovine Spongiform Encephalopatly

The emergence of bovine spongiform encephalopathy (BSE) in 1985/1986 in Great Britain and the postulated aetiological relationship to scrapie suggested that there might also be a correlation between scrapie and human TSE, such as CJD or GSS. Although there are no scientifically proven data to strengthen such an assumption, there are clinical and pathohistological similarities between human and animal TSE which seem to support such a theory. This was further suggested by the appearance of new variant CJD (vCJD), which is believed to be caused by BSE and thus originally by scrapie. All epidemiological data, however, contradict those speculations. The very low incidence of CJD, which is in the range of 10^{-6}, is about the same in all countries whether scrapie is endemic there or not.

References

1   Bendheim PE: Natural scrapie in sheep. A review. Isr J Vet Med 1993;48.
2   Detwiler LA: Scrapie. Rev Sci Tech Off Int Epiz 1992;11:491–537.
3   Detwiler LA, Jenny AL, Rubenstein R, Wineland NE: Scrapie: A review. Sheep Goat Res J 1996; 12:111–131.
4   VanderGaast NE, Miller BS: A Brief Review of Scrapie. Iowa State University Veterinarian Vol 52, No. 1989.
5   Aguzzi A: Molekulare Pathogenese der spongiformen Enzephalopathien. Verh Dt Ges Pathol 1997; 81:35–47.
6   Esmonde TFG, Will RG: Transmissible spongiform encephalopathies and human neurodegenerative disease. Br J Hosp Med 1993;49:400–404.
7   Bradley R: Bovine spongiform encephalopathy and its relationship to the new variant form of Creutzfeldt-Jakob Disease; in: Rabenau HF, Cinatl J, Doerr HW (eds): Prions. A Challenge for Science, Medicine and Public Health System. Contrib Microbiol. Basel, Karger, 2001, vol 7, pp 105–144.
8   Hoinville IJ: A review of the epidemiology of scrapie in sheep. Rev Sci Tech Off Int Epiz 1996;15: 827–852.
9   Hunter N: PrP genetics in sheep and the applications for scrapie and BSE. Trends Microbiol 1997; 5:331–334.
10 Kimberlin RH: Review of scrapie-like diseases; in Bradley et al. (eds): Sub-acute Spongiform Encephalopathies. 1986, pp 1–9.
11 Taylor DM, Fernie K, McConnell I, Steele PJ: Survival of scrapie agent after exposure to sodium dodecyl sulphate and heat. Veterinary Microbiology 1999;67:13–16.

Prof. Dr. Oskar-Rüger Kaaden, Veterinärstrasse 13, D–80539 Munich (Germany)
Tel. +49 89 2180 2528, Fax +49 89 2180 2597, E-Mail o.kaaden@lrz.uni-muenchen-de

Rabenau HF, Cinatl J, Doerr HW (eds): Prions. A Challenge for Science,
Medicine and Public Health System. Contrib Microb. Basel, Karger, 2001, vol 7, pp 151–159

••••••••••••••••••••••••

The Challenge for the
Public Health System

Philip D. Minor

National Institute for Biological Standards and Control, South Mimms,
Potters Bar, UK

Creutzfeldt-Jakob disease (CJD) and the transmissible spongiform encephalopathies (TSEs) of animals are characterized by very long asymptomatic
incubation periods. There is currently no applicable preclinical diagnostic test,
and once symptoms develop there is as yet no treatment. If a suspected
transmission route is blocked, no effect on numbers of cases can be expected
for many years; for example while the ban on feeding ruminant-derived protein
to cattle was introduced in July 1988 to stop the transmission of bovine
spongiform encephalopathy (BSE), the peak incidence of disease was towards
the end of 1992 and the beginning of 1993. These are the challenges of the
TSEs to public health and medical practice.

Properties of Transmissible Spongiform Encephalopathies

While there is controversy over almost every aspect of the details of TSEs,
there is a consensus on a number of general features. Firstly, TSEs involve
a slowly progressing non-inflammatory neurological degeneration, and once
symptoms develop they are invariably fatal. The disease has features of both
genetic and infectious disease; TSEs are clearly transmissible, as was shown
in the 1940s when a vaccine against the sheep disease Louping Ill, prepared
from the spinal cords of infected sheep, transmitted scrapie at high frequency
[1]. On the other hand, there is equally clearly a strong genetic element to the
diseases; CJD can occur in humans in a familial form with 100% penetrance,
and flocks of sheep of certain genotypes are particularly susceptible to the
development of scrapie [2]. Despite this, the diseases are immunologically

Table 1. TSEs

Disease	Natural host
Scrapie	sheep and goats
Transmissible mink encephalopathy	mink
Chronic wasting disease	mule, deer and elk
Bovine spongiform encephalopathy	cattle
Feline spongiform encephalopathy	cats
Exotic ungulate encephalopathy	Kudu, nyala, oryx, gemsbok, eland
Kuru	humans – Fore tribe
Creutzfeldt-Jakob disease	humans
Gerstmann-Sträussler-Schinker syndrome	humans
Fatal familial insomnia	humans

silent; no antibody or cellular immune response has been observed in animals or humans incubating the disease and the clinical presentation is of an encephalopathy with no inflammatory response rather than an encephalitis. There is as yet no reliable surrogate marker for infection, and the only satisfactory assay system still involves measurement of infectivity in vivo. Infectivity is extremely difficult to destroy completely, being refractory to conventional autoclaving techniques and most disinfectants; autoclaving in the presence of 1 M sodium hydroxide or soaking in hypochlorite with adequate levels of free chlorine are currently thought to be the only certain methods of disinfection [3, this volume]. Disease is associated to a greater or lesser extent with the deposition of the normal cellular protein PrP^C in an abnormal isoform, designated PrP^{Sc}. The ratio between infectivity and PrP^{Sc} deposition depends on the strain of the agent concerned, of which there may be more than twenty in the case of sheep scrapie. Each strain has its own distinct and inheritable properties, including pathogenesis and susceptibility to inactivation procedures. Finally, infectivity is found in the CNS, especially the brain, but is not confined to it, also appearing in lymphoid tissues, depending on the strain and model considered.

Types of Transmissible Spongiform Encephalopathies of Concern

The commonly considered TSEs are listed in table 1. Kuru was transmitted as a result of certain funerary practices of the Fore tribe of Papua New Guinea, and BSE was transmitted at the height of the UK epidemic by feeding the rendered carcasses of infected cattle to cattle. Neither of these mechanisms is a natural means of maintaining infection in populations as conventionally understood. The only TSEs shown in table 1 that are known to be naturally self-sustaining are scrapie of sheep and chronic wasting disease of mule deer and elk. The remainder present as diseases restricted to the affected animal or human unless experimentally transmitted by artificial routes.

It is impossible to be sure that an infectious disease will not change its host range or other properties; for example, BSE is believed by most workers to have originated from sheep scrapie although it is clearly distinct from known strains in its properties. However, there is no evidence that classical scrapie of sheep and goats is capable of transmission to humans as CJD. The incidence of CJD does not follow exposure to sheep-derived materials or the presence of scrapie in a country, being surprisingly uniform throughout the world at about 1 case per million head of population per year [4]. Not all TSEs are therefore of major proven importance to human health. The cause of sporadic CJD is not known, although it is believed to be non-infectious, while familial CJD and Gerstmann-Sträussler-Schinker syndrome are both genetic in origin rather than transmissible. Both forms are relatively rare, so that the problems they pose are those of a distressing clinical illness rather than a major public health problem. However, one or more of these diseases was responsible for the occurrence of iatrogenic CJD transmitted by surgical instruments, corneal transplantation, dura mater or hormones derived from human cadaveric pituitaries (see table 5 in Zerr and Poser [5], in this volume).

The most effective strategy to prevent iatrogenic transmission has involved alternative or highly selective sourcing. For example, in the United Kingdom dura mater is no longer used, having been replaced in neurosurgical procedures by autologous fascia, and growth hormone is now produced by recombinant DNA technology rather than from cadaveric pituitaries. However, dura mater continues to be used in the USA and there is no substitute for human cornea. Safety is sought by careful examination of the donors to ensure freedom from neurological disease, and in the case of dura mater, the use of unpooled preparations of raw materials then subjected to particular processes.

The first case of the variant form of CJD termed vCJD was identified in 1995, and all cases of vCJD examined so far have the same strain characteristics as BSE and strains derived from BSE either experimentally by laboratory passage or by feeding material from infected cattle to domestic cats or zoo animals [6, 7]. The conclusion that the agent of BSE and of vCJD are the same seems inescapable, and the most likely explanation for the occurrence of vCJD is that it resulted from exposure of the victims to BSE by some route, probably oral. Thus BSE is believed to be transmissible to humans in a way in which scrapie is not, and moreover human exposure to beef and beef products in various forms is far greater than the corresponding exposure to ovine material; consequently, BSE is a major concern for human health.

By the same token, vCJD is a major concern as it is a new disease, and would almost certainly be more readily transmitted to humans than BSE by the same route.

Finally, in the United Kingdom and elsewhere during the BSE epidemic, ruminant derived protein was used as a feed supplement for sheep. BSE can be transmitted to sheep experimentally by the oral route where the tissue distribution has been shown to be the same as in classical scrapie. Consequently, the theoretical possibility exists that BSE has been introduced into the general sheep population, where it might be expected to be self-sustaining, as is scrapie, while retaining the strain characteristics of BSE, including the ability to transmit to humans. Scrapie is now considered by some as a potential cause for concern, although transmission of BSE to sheep in an agricultural rather than experimental setting has not been demonstrated.

Minimizing the Risks of Transmission of Spongiform Encephalopathies by Medicinal Products

In 1991, the Committee for Proprietary Medicinal Products (CPMP) of the European Union issued a Note for Guidance on Minimizing the Risks of Transmission of BSE by Biological Medicinal Products for Human Use. After the first public recognition of vCJD in 1996, the document was revised somewhat, although the principles on which the first document was based were considered to be sound. The strategy chiefly involved sourcing from animals not at risk from BSE because they came from BSE-free countries, and secondly from tissues expected to have little or no risk of infectivity. Thus brain was regarded as a potentially highly dangerous tissue while serum was not, based on early studies of scrapie in sheep [8]. In addition, while it was recognized that the agents are difficult to destroy, it was considered that the manufacturing process could contribute to safety by removal or inactivation of infectivity. Studies to demonstrate the effect of the process are however difficult to perform convincingly. The revised document also indicates that other TSEs, specifically those of ruminants are of concern.

Objections to the approach taken could include the degree of certainty with which any of the strategies can be documented. The definition of a country as BSE free has been given by the Organisation Internationale des

Epizooties (OIE) and the Scientific Steering Committee (SSC) of DG24 of the European Commission, but there are major debates surrounding the adequacy of surveillance and the control of feeding practices required. There are also problems associated with the movement of cattle and their traceability; for example the producers of Brain-Heart Infusion broth used in bacterial culture were unable to identify the country of origin of the source animals with certainty. Similarly, the classification of tissues used can pose problems where there is a possibility that a low risk tissue such as bone can be contaminated by a high risk tissue such as brain or spinal cord. The difficulties of relying on the manufacturing process unless it involves total chemical degradation are obvious from the hardy nature of the agents. Consequently, products have tended to be considered on a case by case basis.

The issues raised by BSE in particular are very broad, covering a variety of foods and medicines, and the responsibilities for regulation are equally widespread. This has led to problems where solutions have been proposed in one area which have caused enormous difficulties in others. An example was the Commission Decision 97/534/EC to ban the use of all specified risk materials (SRMs) from use for any purpose. SRMs included the skulls and spinal cords of cattle, sheep and goats as well as certain lymphoid tissues of sheep and goats. This approach made good sense from the agricultural point of view, and would have bypassed most of the difficulties associated with uncertainties of geographical sourcing, but it caused great difficulty for the pharmaceutical industry which uses large amounts of tallow derivatives such as magnesium stearate. Tallow derivaties are in fact produced by harsh chemical treatments capable of inactivating TSEs and are therefore considered safe despite the possible inclusion of SRMs in their starting materials. The regulatory approach has evolved subsequently, and continues to do so.

Implications of the Variant Form of Creutzfeldt-Jakob Disease: Blood and Blood Products

Blood and its cellular components are used in a number of life-saving medical interventions, and blood-derived proteins, such as albumin, immunoglobulin and clotting factors such as factor VIII and factor IX are used to treat patients who would otherwise have a very poor life expectancy. Studies on the pathogenesis of scrapie in sheep [8] and in a variety of animal models lead to the conclusion that there is a stage of the disease outside the nervous system, specifically involving the immune system although the cell type remains controversial [9, 10]. In a number of animal models and in scrapie itself, infectivity has been identified in blood [11]. Experiments to detect

CJD infectivity in human blood have been equivocal and epidemiological studies of recipients of blood and blood products have failed to identify any demonstrable risk [11]. It remains possible that infectivity is found in the blood of classical CJD victims, but at such a low level, or in so few donors that the effect is undetectable by epidemiological or experimental methods. Those at risk of CJD because of a family history or previous medical treatment are therefore excluded from donation in both Europe and the USA. However, there was until recently some disagreement about the action to be taken should it be found after donation that an individual was in fact at risk or developed classical CJD. In the USA, the decision taken was that all products to which such a donor contributed should be recalled as a precautionary measure. While very few donors were involved every year, this caused major disruptions of supply, because a single donation contributes to many products. Up to 20% of some products were the subject of recall. In Europe, the reasoning was that there was no evidence that blood or blood products have ever transmitted infection, so that any risk must be extremely small, that in view of the assumed prolonged incubation period of the disease if there were infectivity in the blood, considerably more is likely to come from asymptomatic than symptomatic individuals, so that withdrawal would not in any case contribute to safety significantly, and that the problems of supplying products would cause clinically unacceptable failures to treat patients. Withdrawal after the event was therefore not undertaken in Europe, and the current USA policy has now been modified to essentially the same approach. However the situation with vCJD could be entirely different. It is known that the abnormal isoform of PrP can be detected more easily outside the CNS in lymphoid tissue in vCJD than in classical CJD so that levels in the blood could be higher. Furthermore estimates of the size of the epidemic to be expected in the UK range up to over 100,000 [12] so that exposure could be far higher than with the classical form. Consequently, it was decided to stop the production of blood products from UK plasma, although the use of UK-sourced blood and cellular components would continue as they could not be supplied from other countries.

This decision creates difficulties for other countries. While it is clear that at present all but three of the cases of vCJD have occurred in the UK, the precise means of infection is not established. It can be argued that anyone resident in the United Kingdom during the period of the BSE epidemic should be excluded from donation wherever they live currently; this in turn leads to the question of what is meant by the term 'resident', which could imply a long period or a 2-week vacation. This issue is unresolved at the time of writing, and arises from lack of knowledge of some of the basic parameters associated with the disease. A second decision made by the United Kingdom government has been to introduce leucodepletion to remove white blood cells from the components used, as a means of reducing infectivity. While the factual basis underlying this decision could be questioned [13], there are likely to be other benefits of leucodepletion which also justify it. The steps taken to counter the possible risk of transmission of vCJD by blood or blood products are very costly. There remains some debate as to whether they are either necessary or effective, a problem which is associated with many aspects of the public health impact of TSEs.

Implications of the Variant Form of Creutzfeldt-Jakob Disease: Other Medical Interventions

The peripheral distribution of the abnormal form of PrP, which is believed by many to be the infectious entity itself raises questions concerning surgical interventions. It is already well known that classical CJD can be transmitted by neurological instruments such as electrodes, and that they cannot be adequately sterilized by standard methods. Only disposable instruments of this type are therefore used. However, if vCJD infectivity is in fact widely disseminated throughout the infected but asymptomatic individual, it implies that instruments used in more general surgery could also be able to transmit infection. Many such instruments are not readily disposable. Investigations to address the problem include studies of sterilizing cycles; for example autoclaving in the presence of $1\,M$ sodium hydroxide is likely to be effective, although its effect on the instruments may be severe.

The Future

Many of the problems surrounding the TSEs come from the absence of a satisfactory preclinical diagnostic which can be used in the short term to establish the size of the problem and the effect of steps taken to solve it. There is a great deal of work in progress to develop diagnostic methods applicable to readily obtained specimens. Most are based on attempts to detect the abnormal form of PrP, ultimately in blood [14]. The approach assumes that PrP^{Sc} and infectivity will be found in the blood of preclinical cases, and will require an assay of very high sensitivity, probably more sensitive than an infectivity assay. A final question concerns the ethics of testing individuals for a disease for which there is no treatment and which is believed to be invariably fatal. Similar ethical problems arose in testing for HIV, and they can be addressed.

A second issue is the possible development of treatments or prophylaxis. Once symptoms associated with neuronal destruction have developed it may be difficult to reverse them, and until symptoms have developed there is currently no evidence of infection. Nonetheless, there are a number of approaches which have been applied in animal models, including the use of dextran sulphate, and the related pentosan sulphate has been proposed as a human treatment [15]. The obstacles are that it is not known what dose by what route, given how soon after infection and for how long will be required to protect a human subject. The potential treatments are also likely to have effects: pentosan sulphate is an anticoagulant which may give problems if taken over a long period. The development of such treatments would clearly be facilitated by the simultaneous development of diagnostics which could monitor their effects on the incubation of CJD development.

In summary, the challenges to public health posed by TSEs are complex, and arise from the nature of the agents concerned and the difficulties of diagnosis, disinfection and treatment they present. Procedures are in place which can be expected to minimize possible transmissions to humans, but the scale of the problem will only be clear when the current efforts to develop a better understanding of the diseases and their diagnosis bear fruit.

References

1   Gordon WS: Advances in veterinary research. Vet Res 1945;58:516–520.
2   Baker HF, Ridley RM: The genetics and transmissibility of human spongiform encephalopathy. Neurodegeneration 1992;1:3–16.
3   Taylor DM: Resistance of transmissible spongiform encephalopathy agents to decontamination; in Rabenau HF, Cinatl J, Doerr HW (eds): Prions. A Challenge for Science, Medicine and Public Health System. Contrib Microbiol. Basel, Karger, 2001, vol 7, pp 58–67.
4   Ridley RM, Baker HF: Variation on a item of Creutzfeldt-Jakob disease: Implications of new cases with a young age at onset. J Gen Virol 1996;77:2895–2904.
5   Zerr I, Poser S: Epidemiology and risk factors of transmissible spongiform encephalopathies in man; in Rabenau HF, Cinatl J, Doerr HW (eds): Prions. A Challenge for Science, Medicine and Public Health System. Contrib Microbiol. Basel, Karger, 2001, vol 7, pp 93–104.
6   Bruce ME, Will RG, Ironside JW, McConnell I, Drummond D, Suttie A, McCandle L, Chree A, Hope J, Birkett C, Cousens S, Frazer H, Bostock CJ: Transmissions to mice indicate that 'new variant' CJD is caused by the BSE agent. Nature 1997;389:498–501.
7   Collinge J, Sidle KCL, Meads J, Ironside J, Hill AF: Molecular analysis of prion strain variation and the aetiology of 'new variant' CJD. Nature 1996;383:685–690.
8   Hadlow WJ, Kennedy RC, Race RE: Natural infection of Suffolk sheep with scrapie virus. J Infect Dis 1982;146:657–664.
9   Fraser H, Bruce ME, Davies D, Farquhar CF, McBride PA: The lymphoreticular system in the pathogenesis of scrapie; in Prusiner SB, Collinge J, Powell J, Anderton B (eds): Prion Diseases of Humans and Animals. Chichester, Horwood, 1992, pp 308–317.
10   Klein MA, Frigg R, Flechsig E, Raeber AJ, Kalinke U, Bluethmann T, Bootz F, Suter M, Zinkernagel RM, Auzzi A: A crucial role for B cells in neuroinvasive scrapie. Nature 1997;390:687–690.

11 Brown P: Can Creutzfeldt-Jakob disease be transmitted by transfusion? Curr Opin Haematol 1995; 2:472–477.
12 Cousens SN, Vynnycky E, Zeidler M, Will RG, Smith PG: Predicting the CJD epidemic in humans. Nature 1977;385:197–198.
13 Brown P, Rohwer RG, Dunston BC, MacAuley C, Gajdusek DC, Drohan WN: The distribution of infectivity in blood components and plasma derivatives in experimental models of transmissible spongiform encephalopathy. Transfusion 1998;38:810–816.
14 Safar J, Wille H, Itri V, Groth D, Serban H, Torchia M, Cohen FE, Prusiner SB: Eight prion strains have PrP^{Sc} molecules with different conformations. Nat Med 1998;4:1157–1165.
15 Diringer H, Ehlers B: Chemoprophylaxis of scrapie in mice. J Gen Virol 1991;72:457–460.

Dr. Philip D. Minor, National Institute for Biological Standards and Control,
Blanche Lane, South Mimms, Potters Bar, EN6 3QG (UK)
Tel. +44 1707 65 47 53; Fax +44 1707 64 67 30; E-Mail pminor@nibsc.ac.uk

Subject Index